# PROTOPLASMA

*An International Journal of Cell Biology*

Edited by
**T. W. Goodwin,** Liverpool
**B. E. S. Gunning,** Canberra
**O. Kiermayer,** Salzburg
**D. J. Morré,** West Lafayette, Ind.
**K. R. Porter,** Boulder, Colo.
**J. Reinert,** Berlin
**E. Schnepf,** Heidelberg
**M. Tazawa,** Tokyo

The Journal publishes original manuscripts in any area of biology of cell—animal, plant, algal, protozoan, fungal—including descriptive and experimental ultrastructure, differentiation, cytoskeletal elements, membranes and membrane biogenesis, cell fractionation (organelle isolation), analysis and intracellular distributions of cell constituents, and papers dealing with new methodologies in these areas. Preference is given to papers with significant content relating to structure-function correlations.

PROTOPLASMA offers best chances for the priority of new results by its short publication time. At the present state the time lag between submission of a paper to the editor and publication—in case of acceptance without changes—is less than five months.

# Cell Biology Monographs

*Volume 11*

*Springer-Verlag*

**Wien New York**

# DNA Methylation
# and Cellular Differentiation

## J. Herbert Taylor

**Springer-Verlag**
**Wien New York**

Prof. Dr. JAMES HERBERT TAYLOR

Institute of Molecular Biophysics,

The Florida State University, Tallahassee, Florida, U.S.A.

With 17 Figures

Library of Congress Cataloging in Publication Data. Taylor, J. Herbert (James Herbert), 1916— . DNA methylation and cellular differentiation. (Cell biology monographs, ISSN 0172-4665; v. 11.) Bibliography: p. Includes index. 1. Deoxyribonucleic acid—Metabolism. 2. Methylation. 3. Cell differentiation. I. Title. II. Series. QP 624.T 39. 1984. 574.87'3282. 83-20239

ISSN 0172-4665

ISBN-13:978-3-7091-8723-4     e-ISBN-13:978-3-7091-8721-0

DOI: 10.1007/978-3-7091-8721-0

# Preface

In 1977 I wrote a grant proposal in which I applied to study developmental patterns in enzymatic methylation of DNA in eukaryotes. One part of the proposal was to assay cells at different embryonic developmental stages for maintenance and *de novo* type methylase activity. With one exception the referees, probably developmental biologists, recommended that the work not be supported because there was no evidence that methylation plays any role in eukaryotic gene regulation. Aside from proving that innovative ideas can seldom be used to successfully compete for grant funds, the skepticism of biologists toward methylation as a regulatory mechanism was, and still is, widespread even among some of those who investigate the problem. That is a healthy situation for all points of view should be brought to bear on a problem of such importance. However, to deny funds to investigate a problem because one has already formed an opinion without evidence is hardly commendable. The great skepticism about the significance of DNA methylation is based in part on the evidence that it is absent or very little used in *Drosophila,* a favorite organism for genetic and developmental studies. There now remains little doubt that methylation of cytosine in certain CpG sites can strikingly affect the transcription of sequences 3′ to the methylated doublet. How this inhibition operates and to what extent it is utilized in cells is still debatable. Furthermore, a mechanism for the inheritance of a pattern of methylation once it is installed is understood and demonstrated in principle. What other functions DNA methylation may have is debatable. Certainly most methylation of CpG sites does not appear to affect transcription and many plants, which have up to one-third of the cytosines methylated, provide an enigma.

This little book is not a review of everything published on DNA methylation. Several short reviews have been published in the last several years (TAYLOR 1979, RAZIN and RIGGS 1980, YUAN 1981, RAZIN and FRIEDMAN 1981, DOERFLER 1981, ADAMS and BURDON 1982, and RIGGS and JONES 1983). The interested reader may consult these reviews for further details. What I have tried to do is to survey the literature and to pick those investigations which are relevant to cellular differentiation, or provide background for understanding the problems. I have then picked a few examples and tried to relate them to a differentiating system and to see how the two fit together. Investigations are few and much remains to be learned, I am sure. Probably the most frustrating aspect of studies of DNA methylation is the paucity of methods for investigating the 5-methyl cytosine in genes before cloning. The methylated cytosine is lost when sequences are cloned and amplified for sequencing, because bacterial hosts for the cloning vectors do not have the maintenance enzymes to reproducibly methylate all or a major fraction of the CpG doublets in DNA. Perhaps that situation points to our best hope for the future. If the maintenance methylases can be characterized, isolated and cloned, we may be able to develop hosts for cloning vectors with the eukaryotic maintenance methylases. However, we are far from that stage today. We must be satisfied with a few restriction enzymes that are sensitive to methylation at CpG or CpC sites and to the sequencing of a few DNA repeated sequences that can be prepared in enough copies for sequencing without amplification. Another recent development with great promise is the sequencing of single copy genes. It involves an adaptation of the Southern blot of a gel after a Maxam-Gilbert type reaction in which the probe is a fragment at one end of the sequence to be determined.

For those who read prefaces and do not like to read books from the end to the beginning, let me state my conclusions very briefly. I have concluded that DNA methylation is an important mechanism of cellular differentiation in vertebrates, although it probably is not in insects such as *Drosophila*. Even in vertebrates, methylation is only a subsidiary mechanism. The primary mechanism is based on a very simple scheme that eukaryotic cells invented hundreds of millions of years ago and cell biologists and geneticists have studied for years, most of them without understanding its real significance. Cells have partitioned their DNA into two pools for replication and packaging, an early replicating pool and a late replicating pool with a short pause between. It is possible that some cells have invented a third or fourth pool, although I doubt that elaboration is either necessary or desirable. In the first pool, replicated in $S_E$, are all of the genes that the cell will need to use in its differentiating processes and functional roles. All other genes replicate in $S_L$, the last half of the S phase, and are maintained in an inactive state. The proteins that sequester genes from transcription are available in $S_L$. The proteins that open up the genes to transcription are available only in $S_E$ and any sequence replicated in $S_E$ is potentially functional when and if the cellular environment is appropriate. This situation makes the control of the time of replication in the cell cycle crucial, but it probably makes little if any difference in what part of $S_E$ or $S_L$ a particular gene replicates. It also means that a mechanism to modify replication origins in a way that is stably inherited is a crucial feature of differentiation. Some replicons must be shifted to replicate in $S_E$, others to $S_L$.

Methylation as a part of differentiation may be a late development in evolution and is only extensively exploited in vertebrates in which it suppresses those genes in a cluster (replicon) that will not be useful in a particular differentiated cell. For example, in a red cell precursor all globin genes will be switched to replicate in $S_E$, but only those will be expressed which are demethylated during a subsequent determination step.

I know that both of these ideas, and particularly the first, will be considered naive and unsupported by all of the evidence by some critics, but I predict that biologists will still be investigating both these phenomena and acquiring astonishing results when the critics and I have passed from the scene.

I thank those who have cooperated with me by sending illustrations; each is credited in the legends of the illustrations except where there are multiple authors and it may not be obvious which one helped. I would also like to give credit to Dr. M. BUSSLINGER and Dr. GERHART RYFFEL, who provided me with valuable information prior to its publication, and to Mrs. MYRNA HURST, who learned to use a word processor to record and revise the manuscript. I wish to acknowledge a one-half year sabbatical leave from Florida State University that enabled me to devote most of my time to the writing of the monograph.

Tallahassee, Florida, October 1983                     J. HERBERT TAYLOR

# Contents

I.  **DNA Methylation and Cell Differentiation: An Overview** . . . . . . . . . . . . . . . 3
    A. Introduction . . . . . . . . . . . . . . . . . . . . . . . . . 3
    B. What New Properties Does Methylation Confer on DNA? . . . . . . . . . 4
    C. The Origin and Maintenance of Methyl Cytosine in DNA . . . . . . . . 7
    D. Differentiation: The Problem Posed . . . . . . . . . . . . . . . . 9
    E. Genome Modifications Which Can Be Associated with Differentiation . . . . . . 12
       1. The Insect Type of Differentiation . . . . . . . . . . . . . . 12
       2. Is the Determined State Reversible? . . . . . . . . . . . . . . 14
       3. The Vertebrate Type of Differentiation . . . . . . . . . . . . 14
       4. Modifications of the Genome . . . . . . . . . . . . . . . . . 15
       5. Mobile Genetic Elements . . . . . . . . . . . . . . . . . . . 16
          a) Examples of Mobile Genetic Elements in Corn . . . . . . . . . 17
          b) The Discovery of a Mutable Gene . . . . . . . . . . . . . . 20

II. **DNA Methylation and Transposable Genetic Elements** . . . . . . . . . . . . . 24
    A. Discovery of Enzymatic Methylation of DNA . . . . . . . . . . . . . 24
    B. The First Demonstrated Role of DNA Methylation: Restriction-Modification Systems
       in Bacteria . . . . . . . . . . . . . . . . . . . . . . . . . 25
    C. Hypotheses for Roles of Methylation in Eukaryotes . . . . . . . . . . 27
    D. Transposable Genetic Elements . . . . . . . . . . . . . . . . . . 32
       1. Flagellar Phase Variation of *Salmonella* . . . . . . . . . . . . 32
       2. Insertion Sequences and Transposons . . . . . . . . . . . . . . 37
          a) Insertion Sequences . . . . . . . . . . . . . . . . . . . 37
          b) Transposons in Bacteria . . . . . . . . . . . . . . . . . . 37
          c) Transposons in Eukaryotes . . . . . . . . . . . . . . . . . 40
             1. A Transposable Element in Yeast, Ty . . . . . . . . . . . 40
             2. Transposable Elements in *Drosophila* . . . . . . . . . . . 41
             3. Retroviruses . . . . . . . . . . . . . . . . . . . . . 42
             4. Alu Family of Sequences in the Human Genome . . . . . . . . 42

III. **Differentiating Systems and Their Methylation Patterns** . . . . . . . . . . . . 46
    A. Differentiation of the Hemopoietic System and Organization of Hemoglobin Genes . 46
    B. Methylation Patterns of Hemoglobin Genes . . . . . . . . . . . . . . 50
       1. A Method for Locating Single CpG Sites by the Use of Two Restriction Endo-
          nucleases Which Are Isoschizomers . . . . . . . . . . . . . . . 50
       2. The Genes of the Beta Cluster Which Function in a Reticulocyte Are Undermethy-
          lated . . . . . . . . . . . . . . . . . . . . . . . . . . . 51
       3. Tissue Specific DNA Methylation Occurs in the Cluster of Rabbit Beta-Like Globin
          Genes . . . . . . . . . . . . . . . . . . . . . . . . . . . 53
       4. The Alpha-Globin Gene Cluster in the Chicken and *Xenopus* . . . . . . 54
    C. Differentiation of Lymphocytes and the Role of Methylation . . . . . . . 58

D. Suppression of Integrated Viral Genomes in Cells by DNA Methylation . . . . . .   63
   1. Small DNA Viruses . . . . . . . . . . . . . . . . . . . . . . . .   63
   2. Inactivation of Retroviral Genomes by DNA Methylation . . . . . . . . .   64
E. The Vitellogenin Genes in *Xenopus* Are Methylated, But Can Be Expressed Without a
   Detectable Change in the Pattern . . . . . . . . . . . . . . . . .   65
F. Integrated Retroviruses Are Methylated Early in Development . . . . . . . .   67
G. Suppression of Metallothionein Genes by Methylation . . . . . . . . . .   68
H. Other Genes That Are Suppressed by Methylation . . . . . . . . . . . .   69
I.  Correlation Between DNase I Sensitivity of Chromatin and Cytosine Methylation . .   70
J.  The HpaII Sites in an Ovalbumin Gene . . . . . . . . . . . . . . .   71
K. The 5′ End of the Rat Albumin Gene Is Undermethylated in Cells in Which It Is
   Expressed . . . . . . . . . . . . . . . . . . . . . . . . .   72
L.  Methylation of Human Growth Hormone and Somatotropin Genes . . . . . . .   73
M. Methylation of Ribosomal Genes . . . . . . . . . . . . . . . . . .   74

IV. **DNA Methylation and the Inactive X Chromosome of Mammals** . . . . . . . .   76
A. A Brief History of X Chromosome Inactivation Studies . . . . . . . . . .   76
B. A New Methylation Model. . . . . . . . . . . . . . . . . . . . .   79

V. **Mechanisms of Suppression by DNA Methylation** . . . . . . . . . . . .   84
A. Expression of the Late Viral Protein of SV-40 (VP-1) Is Reduced by Methylation at One
   HpaII Site . . . . . . . . . . . . . . . . . . . . . . . . .   84
B. Transcription of a Cloned Adenovirus Gene Is Inhibited by *in vitro* Methylation . .   87
C. Transcription of a Cloned Human Gamma Globin Gene Is Inhibited by Methylation at
   the 5′ Region Flanking the Structural Gene . . . . . . . . . . . . .   88
D. The Mechanism by Which Methylated CpG Sites Inhibit Transcription . . . . . .   93
E. How Are Methylation Patterns Established and Maintained? . . . . . . . .   94
F. The Specificity of Methylases and the Maintenance of Methylation Patterns . . . .   94
   1. Prokaryotic Type II Methylases . . . . . . . . . . . . . . . .   94
   2. Type I and Type III Methylases . . . . . . . . . . . . . . . .   95
G. Properties of Eukaryotic Methylases . . . . . . . . . . . . . . . .   96
H. The Maintenance of Methylation Patterns Imposed *in vitro* . . . . . . . . .   97
I.  Deletion of Methylation Patterns During Differentiation. . . . . . . . . .  100

VI. **Evolution, Stability and Regulation of Methylation Patterns** . . . . . . . . .  104
A. The Evolutionary Aspects of DNA Methylation . . . . . . . . . . . .  104
   1. 5-Methylcytosine Residues Are Hotspots for Mutation in Bacteria . . . . .  104
   2. Are 5-Methylcytosine Residues Hotspots for Mutation in Eukaryotes? . . . . .  105
B. DNA Methylation and Repair . . . . . . . . . . . . . . . . . .  107
C. How Stable Is a Pattern of Methylation? . . . . . . . . . . . . . .  108
D. Overview on the Role of DNA Methylation . . . . . . . . . . . . . .  111
E. An Hypothesis for the Control of Methylation Patterns . . . . . . . . . .  114
F. A Pyramid of Controls in Vertebrate Cells . . . . . . . . . . . . . .  117

**References** . . . . . . . . . . . . . . . . . . . . . . . . . . .  120

**Subject Index** . . . . . . . . . . . . . . . . . . . . . . . . .  134

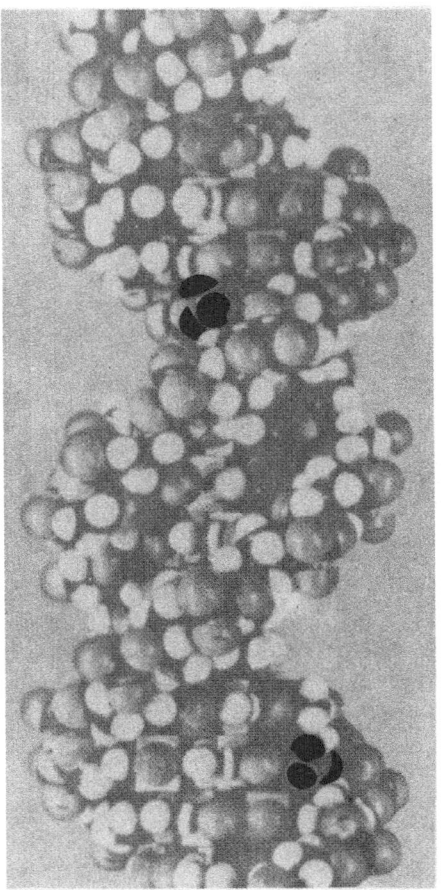

Fig. 1. Methylated DNA. Location of the methyl group of 5-methyl-2'-deoxycytidine in the B-form of a DNA helix. The methyl group projects into the major groove and occupies a position analogous to the methyl group of thymidine. Therefore, the conversion of deoxycytidine in DNA to 5-methyl-deoxycytidine makes the G : C base pair resemble an A : T base pair and a molecule binding to DNA at that point should encounter the difference. (From RAZIN and RIGGS; copyright 1980 by the American Association for the Advancement of Science)

# I. DNA Methylation and Cell Differentiation: An Overview

## A. Introduction

The problem of cellular differentiation has renewed interest for molecular biologists who are studying the modifications of DNA. Three modifications are getting much of the attention. One is a structural modification of the one dimensional message by the insertion of transposable genetic elements. Another is a base modification by methylation which does not alter the one-dimensional coding properties, but may change the affinity for other molecules and thereby transcription or readout of the code. The third is a new three dimensional structural form called Z-DNA which is related to the base sequence, the modification by methylation, the torsional properties and the molecular environment. This change would affect affinity for other molecules and could also affect the readout of the code.

We will be concerned here primarily with modification by methylation, but we can not ignore the other modifications and hope to understand the roles of methylation. All three modifications are reversible, but the insertion of transposable genetic elements requires a precise removal of the insert to restore the original sequence and these appear to be rare events. The modification by methylation is probably removed by dilution during semiconservative replication in the absence of methylase activity at certain sites. The conversion from the Z form back to the more usual B form requires a change in the molecular environment, but could be influenced by methylation and the torsional state of the chains. It is unlikely to persist through replication, but could reform afterwards if the conditions were favorable.

The modification of major concern to us is the methylation of cytosine at the fifth carbon in the DNA polymer after replication. In the duplex these methyl groups project into the major groove of the double helix (Fig. 1) and make the cytosine of a C:G base pair have a conformation more like a thymine residue of a T:A base pair than the usual cytosine. The maintenance of a methylation pattern is assumed to operate by an enzyme that inserts the methyl group on a new chain symmetrically to the one

1*

retained on the parental chain. A highly specific enzyme could make the replacement almost as efficient as replication of a base sequence and the maintenance of patterns observed so far suggests a high fidelity. At some time when a new pattern is initiated, the enzyme would presumably require highly accurate sequence recognition to avoid complete methylation of all cytosines, which of course does not occur. Since most cells are unlikely to have demethylating enzymes, it is generally assumed that loss or change of a pattern occurs by a failure to maintain the pattern at specific sites when replication is occurring. After two rounds of replication, only two of the four segments of the DNA helix will have the hemimethylated DNA with a pattern for an enzyme to act upon. The pattern would be lost permanently from the other two segments (see section C, below). Here also sequence specificity would have to play a role in the positioning of a site specific inhibitor of the enzyme, which normally restores the modification on the new chains after replication, since the methylase activity must always be present for methylation of other sites after replication.

## B. What New Properties Does Methylation Confer on DNA?

In 1948 HOTCHKISS (HOTCHKISS 1948) discovered that calf thymus DNA contained, in addition to the four principal purine and pyrimidine bases, a small amount of 5-methylcytosine (Fig. 2 shows a nucleotide and related molecular), and a few years later WYATT (1951) found higher amounts in DNA from wheat germ. In 1955 DUNN and SMITH reported another minor base, N6-methyladenine ($N^6A$), in bacterial DNA. When the enzymatic mechanisms of DNA replication were being investigated by KORNBERG and associates (BESSMAN *et al.* 1958), several analogs of the four principal bases were prepared as precursors for tests of incorporation in their in vitro system. Among these was 5-methyl-2'-deoxycytosine triphosphate. The nucleotide was incorporated into DNA in the in vitro system in place of cytosine at rates comparable to the four principal bases. Both of these modified bases form base pairs with little effect on the double helix and both are nearly as stable as the four principal bases in DNA. 5-Methylcytosine (5 mC) substitutes completely for cytosine in the *Xanthomonas* phage, Xp 12 (Kuo *et al.* 1968). Before incorporation into the phage DNA the deoxycytidylate of the host cell is converted to 5-methyl-deoxycytidylate by an enzyme designated deoxycytidylate methyl transferase. The methyl group is derived from tetrahydrofolate which is converted to dihydrofolate in the reaction (FENG *et al.* 1978). The phage nucleotide containing 5mC is synthesized in a manner that resembles the synthesis of thymidylate from deoxyuridylate by the ubiquitous enzyme, thymidylate synthetase. The DNA with 5mC substituted for cytosine has a lower buoyant density in CsCl than predicted from other DNA with the same percentage of adenine plus thymine, and the thermal stability is increased. The melting temperature is reported to be 83.2 degrees C in 0.012 M Na$^+$, which is the highest for any naturally occurring DNA and 6.1 degrees higher than regular DNA with the same percentage of adenine and thymine (EHRLICH *et al.*1975). The small amount of 5mC (about 0.5 mole percent) in most DNA from animal cells would be expected to have very little effect on these properties, but in some higher plants where the average 5mC content can be as high as 6–7 mole percent, *i.e.,* every third cytosine modified, the effect could be significant. Considering the probability that the 5mC is not uniformly distributed the effect on the physical properties could be

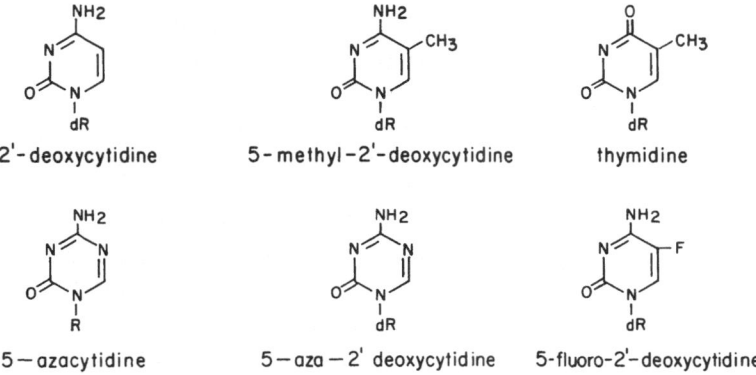

Fig. 2. The tree pyrimidine nucleosides found in DNA and three analogs which can inhibit methylation of cytosine in DNA. 5-azacytidine is not found in DNA, but 5-aza-2'-deoxycytidine diphosphate can be derived from the enzymatic reduction of 5-azacytidine diphosphate and the triphosphate is a precursor for the replacement of 2'-deoxycytidine or 5-methyl-2'-deoxycytidine. 5-fluoro-2'-deoxycytidine can also replace the same nucleosides and the enzymatic methylation should be blocked at the 5 position by the fluorine which replaces hydrogen

dramatic. In addition, BEHE *et al.* (1982) reported that methylation of the synthetic polynucleotide poly (dG:dC)·poly dG:dC) can have a striking effect on its transition from the B form to the Z form. The unmethylated polymer can be converted to the Z form only in a high salt concentration, but the methylated polymer can be converted at a concentration approaching the expected physiological conditions in cells. The polymers used in the experiments were synthesized from nucleotides containing the methylated base. However, the methylated DNA from cells, with the exception of that from the phage Xp 12 mentioned above, is modified by methylation of the polymer after replication by one or more of the class of enzymes called methyl transferases or what we shall usually refer to as DNA methylase.

MANDEL and BOREK (1963) described an enzymatic activity which transferred the methyl group from S-adenosylmethionine (SAM) to polymeric RNA. This discovery initiated a number of studies on the in vitro methylation of transfer RNAs and other types of RNA which is outside the area of our discussion. The methylation of DNA was demonstrated in vitro by GOLD *et al.* (1963 a, b) and by GOLD and HURWITZ (1963) in extracts from bacteria. The donor of the methyl group as for the RNA methyl transferase was SAM. After the modification-restriction system of some phages which infect *E. coli* was shown to be due to methylation of the DNA (LINN and ARBER 1968 and MESELSON and YUAN 1968), the significance of the methyl group for controlling the binding of proteins to DNA became apparent. The methylase, now called type I, binds to DNA which is unmethylated at specific recognition sites, but the reaction is a slow one. However, once bound the enzyme is not released and depending on intracellular conditions may either methylate the DNA or restrict (cleave) it. On the other hand if the DNA is hemimethylated (methylated on one strand at the recognition site), the enzyme rapidly methylates the other chain, but will not restrict (YUAN 1981). The demonstration a few years later that type II restriction endonucleases were also inhibited by methylation of either adenine or cytosine at specific sites which varied for each restriction enzyme (SMITH and WILCOX 1970, and KELLY and SMITH 1970) further emphasized the role of the methyl group on these two bases in regulating the binding as well as the catalytic activity of specific proteins at specific sites on double-stranded DNA. Models of the double helix show that the methyl groups of both 5mC and $N^6A$ project into the large groove (Fig. 1). In the case of 5mC the methyl group at a binding site probably resembles that of thymine as shown by the experiment described below.

An experiment on the binding of *lac* repressor to its specific site on the *E. coli* DNA illustrates this change in affinity in a dramatic way. FISHER and CARUTHERS (1979) studied the affinity of *lac* repressor protein which functions in the regulation of the lactose operon in *E. coli*. Of the 26 nucleotide pairs in the *E. coli* chromosome which constitute the binding site for the *lac* repressor, the A:T pair at position 13 (Fig. 3) is crucial. Mutations with a substitution of a G:C base pair at that position are constitutive for the enzymes coded by the *lac* operon. The binding is insufficient to block expression of the genes even in the absence of lactose. In vitro experiments in which a uracil residue was substituted for the thymine showed that the affinity for the *lac* repressor was dramatically reduced. A tenfold reduction in the stability of the repressor-operator complex was observed. The repressor-operator interaction was assumed to be hydrophobic and consistent with that idea they found that bromouracil substituted for thymine produced a slight reduction in the stability of the complex. In a transversion, in which the adenine-thymine pair was changed to thymine-adenine, there was a seven fold reduction in the stability of the complex. This suggests that the hydrophobic interaction is positionally specific. When 5mC was substituted for cytosine in position 13, the stability of the complex was equal to that of the standard wild type with the A:T base pair at that position. These reconstructions of the site demonstrate that the conversion of a G:C base pair to G:5mC by methylation can have a functional effect similar to a mutation in which the base pair G:C is converted to A:T. However, by the dilution of the methylated base in replication without the action of a methylase, the mutant phenotype would return. We may conclude that the methyl group projecting into the large groove of DNA can make a critical difference in the binding of various specific proteins to double-stranded DNA.

$$5'\text{-T-G-T-G-G-A-A-T-T-G-T-G-}\overline{\text{A}}\text{-G-C-}\overset{\downarrow}{\text{G}}\text{-G-A-T-A-A-C-A-A-T-T-}3'$$
$$\text{-A-C-A-C-C-T-T-A-A-C-A-C-}\underline{\text{T}}\text{-C-G-C-C-T-A-T-T-G-T-T-A-A-}$$

Fig. 3. The *lac* operator DNA sequence from the *E. coli* at which the *lac* repressor protein binds. The arrow indicates the axis of the twofold symmetry in the region and the line above and below the sequence indicates a base pair which when changed by mutation from A:T to G:C has a marked effect on binding of the repressor. (Adapted from FISHER and CARUTHERS, 1979)

## C. The Origin and Maintenance of Methyl Cytosine in DNA

Since the methylcytosine content of DNA varies among species and among different fractions of DNA in the same genome, it could influence the types of proteins which interact with the DNA. The effects of the proteins on the activities of the DNA could in turn change the phenotype. Therefore, we shall examine briefly the ways in which DNA can be modified and how such modifications could be inherited from cell generation to cell generation. An important consideration will also be the stability of the pattern once it is established. All DNA appears to be synthesized with deoxycytidine triphosphate as a precursor of the deoxycytidine nucleotides, with the exception of DNA in a few bacterial phages. Therefore, the pattern of methylation must be imposed by enzymes that act post-replicatively. As first indicated by the catalytic behaviour of the Type I methylating enzymes of bacteria, there is a basis for the hypothesis that hemimethylated DNA might react with methylases in a unique way. Early experiments indicated that cytosine was the only site of enzymatic, post-replicative methylation in the DNA of most eukaryotes. In animals more than 90% of the 5mC found in DNA is in the dinucleotide 5′ pCpG (DOSKOCIL and SORM 1962, GRIPPO *et al.* 1968), which we will refer to as the CpG doublet. If the DNA is methylated on both chains at CpG doublets, the two new helices after replication will be hemimethylated as shown in Fig. 4. An enzyme, with a preferential affinity for the hemimethylated site and an efficient catalytic activity for adding a methyl group to the symmetrical cytosine, could maintain the pattern with a high fidelity from one cell generation to the next. We will designate such enzymes "maintenance enzymes" (HOLLIDAY and PUGH 1975, and TAYLOR 1979). One of the crucial questions will be, do such enzymes exist in replicating cells and if so how faithfully are the patterns maintained?

If the control of DNA function is related to the pattern of 5mC in eukaryotes, there must be enzymes which initiate the patterns. In bacteria all of the methylases so far known can function to initiate a pattern by reacting

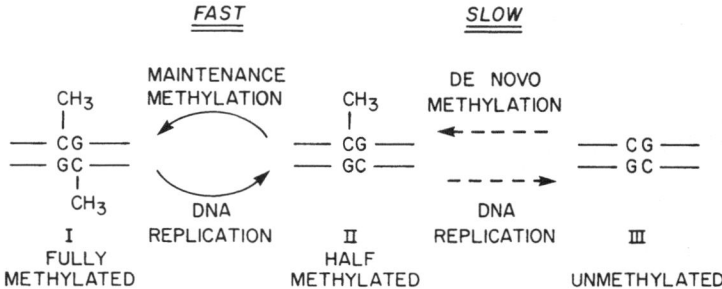

Fig. 4. Three possible states of DNA methylation: (I) fully methylated, (II) half methylated or hemimethylated and (III) unmethylated. Fully methylated DNA would be converted to hemimethylated DNA by ordinary replication which does not utilize 5-methylcytosine containing nucleotides as precursors. However, if a maintenance methylase is available, the hemimethylated DNA will be converted to a fully methylated form before the next replication. If for any reason this does not happen, both chains will be unmethylated after another replication. Maintenance methylase can not restore a fully methylated pattern unless one strand is already methylated. (From RAZIN and RIGGS; copyright 1980 by American Association for the Advancement of Science)

with specific recognition sites, many of which are now known. There is usually a corresponding endonuclease which recognizes the same sequence and cleaves the DNA unless it is methylated on at least one chain. Such endonucleases, which are called restriction enzymes, are rare if they occur at all in eukaryotes. Furthermore, the methylatable sites for a particular enzyme are very efficiently methylated in prokaryotes (bacteria), while in eukaryotes the sites are incompletely methylated. However, methylation patterns appear to be consistent for a particular type of cell at a particular stage in development or differentiation as will be illustrated later on in this monograph.

Methylases which initiate new patterns of methylation have been referred to as de novo (RAZIN and FRIEDMAN 1981) or initiation type methylases (TAYLOR 1979). In prokaryotes, both functions, initiation and maintenance, may be carried out by the same enzyme complex. Not enough is yet known about the methylases of eukaryotic cells to decide this question. However, there is considerable evidence that maintenance of a pattern does occur once it is established and there is circumstantial evidence that some patterns have to be initiated at certain stages in development. Research on the properties of eukaryotic methylases is still in a primitive state in spite of a few interesting and provocative reports (GRUENBAUM *et al.* 1982). Fig. 4 shows schematically how initiation, maintenance and deletion of a pattern might occur.

The other question that will require an answer is how methylation patterns at specific sites within a gene can be changed while similar sites in an adjacent gene are maintained. Since there is only one unconfirmed report of a demethylating enzyme which acts on polymeric DNA (GJERSET and MARTIN 1982); it seems too early to conclude that such enzymes operate to eliminate methylation at specific sequences. It is more likely that methylation is lost by dilution during replication, but the question of a mechanism which would allow specificity for such a dilution remains unanswered. One possibility is a site specific inhibitor, but would cells code enough inhibitors for each individual gene that must be changed to an unmethylated state? Since that seems unlikely, a smaller number of inhibitors, that bind to families of similar sequences and inhibit methylation downstream to the next binding sequence while replication proceeds through two rounds would effectively eliminate the pattern from both strands on two of the four daughter chromosomes or more specifically on certain sequences of those chromosomes (Fig. 4). Available evidence suggests that changes in methylation patterns, either the loss or gain, are associated with cellular proliferation and therefore, with DNA replication.

## D. Differentiation: The Problem Posed

The problem has always been to explain how a single cell, the zygote, with a single genotype can produce a highly organized arrangement of cells with the many phenotypes that make up the body of the individual eukaryotic organism. As long as biologists have accepted the cell theory and begun to understand how cells reproduce and segregate into functional organs, the origin of this diversity has puzzled and intrigued them. The first notably philosophical solution was suggested by WEISMANN (1892) in the last century when he proposed that the genetic determinants were segregated to the different cells in some systematic fashion. With the advent of the chromosome theory of inheritance and the concept of the genes which make up a chromosome, there was a physical basis for this concept. However, as the knowledge of chromosomes developed, the evidence indicated that with few exceptions the chromosomes remained morphologically the same in all of the cells of an individual. There were some notable exceptions such as the discovery that the germline and somatic cells had a different chromosome complement in a few cases. For example, during the early cleavage stages in the embryo of the parasitic worm, *Ascaris,* there is an elimination of parts of the chromosomes in cells that will form the somatic organs. However, the full complement is retained and passed on in the cells of the germline. In some insects, notably certain species of dipterans, there is an elimination of whole chromosomes from the cells that will differentiate into somatic tissues. Of course, this could only account for a difference between the germline and the somatic cells, but polyploidy and polyteny were also discovered in a wider variety of cells. Considerable discussion of the possible significance of these changes can be found in the literature before 1950. The multiplication of all of the genes could not give rise to variations among cells but it was argued that a quantitative difference could possibly arise if there should be a differential replication of the genes in various cells. These discussions were largely laid to rest by two types of experiments and observations beginning in the early 1950's.

BEERMAN (1952, 1962) began publishing the results of systematic developmental studies of the polytene chromosomes of the dipteran insects. The puffs at specific loci which had been viewed as the morphological basis for the presumed differences among the chromosomes in different tissues, were shown to be localized changes in bands of the polytene chromosomes associated with different developmental stages. After puffing or changing in appearance the chromosomes often returned to the original condition in a later of development. A change in the functional state of the genes could occur without a necessary change in the quantity of the DNA.

BRIGGS and KING (1952) began a series of experiments with frog eggs in which nuclei from the cells of blastula, and eventually later stages in development, were injected into enucleated eggs. The nuclei appeared to be capable of reinitiating normal development even when taken from late stages in embryos. However, the efficiency of the nuclei in reinitiating normal development decreased with age of the embryo. These studies indicated that the changes which were presumed to occur during differentiation were largely reversible. GURDON and his co-workers (GURDON 1962 a, b; GURDON and LASKEY 1970) continued these experiments with *Xenopus,* the South African frog. Most of the experiments confirmed the view that whatever changes occur during development and differentiation are reversible. While it is true that many nuclei fail to function to reinitiate development when transplanted to a frog egg, the failure could be attributed to deficiencies other than a loss or permanent change in some of the genes. However, McKINNELL (1978), who has written a monograph on the problem, maintains that there may be non-reversible changes in the genotype during differentiation which such transplantation studies have not revealed. A more recent study of the ability of erythrocyte nuclei to support continued development when transferred to eggs of the frog, *Rana pipiens,* also leaves the question of irreversible changes of these nuclei open (DI BERARDINO and HOFFNER 1983). When erythrocyte nuclei were introduced into enucleated eggs only 7% formed partially cleaved blastulas. Most of the blastulas were used as donors for recycling the nuclei, *i.e.,* injecting them into another enucleated egg. Although 49% of these eggs developed into blastula, only a few reached early stages of gastrulation. However, when the erthrocyte nuclei were injected into oocytes at the first meiotic metaphase and the oocytes allowed to mature in vitro the results were different. After 24 hours the matured oocytes were activated by pricking with a glass needle and the nucleus removed at second meiotic metaphase, the erythrocyte nuclei participated in development. Sixteen percent reached the blastula and a few went as far as neurula stages. The recycled nuclei from some of the blastula produced 63% blastula and 10% swimming larva, but when the tadpoles reached the feeding stage all died.

Another system that may give even more information on mechanisms of differentiation has been developed by MINTZ and her associates over the last several years (MINTZ 1978, MINTZ *et al.* 1978, MINTZ and CRONMILLER 1981, and WAGNER *et al.* 1981). The unusual tumors called teratocarcinomas have been maintained by transplants and grown in culture for a long time, but recently MINTZ developed a new cell line with a normal set of mouse chromosomes which is developmentally totipotent, *i.e.,* it can under appropriate conditions when transplanted into early embryos (blastocysts) form any tissue of the adult mouse, including cells of both soma and germline. It loses its malignant characteristics after passage through the embryo and participates in normal development along with other cells of the embryo into which it is transplanted. This demonstrates that certain traits (malignancy at least) acquired in development are reversible, although other traits may not be, as we will see when the lymphocytes are studied. The genomes of these latter cells change during maturation into plasma cells or when arrested by tumorigenic transformation during differentiation. Other transformed cells will not function like those of teratocarcinomas and most of them become polyploid or aneuploid during culture in vitro.

The special cell line METT1 (Mouse Euploid Totipotent Teratocarcinoma) isolated by MINTZ and CRONMILLER (1981) is unusual among teratocarcinomas in that it grows in culture without a feeder layer of cells and after numerous passages, including cloning, can still participate in normal development when transplanted into blastocysts. Mutants can be induced in the cell line or it can be transformed by the insertion of cloned genes, for example, thymidine kinase and human $\beta$-globin in a plasmid vector. After such genetic manipulations the totipotency of METT-1 cells is retained. This system provides a powerful technique for modifying genes in a variety of ways and testing the effects on expression during differentiation. One limitation is that the transformed cells may carry the genes in unusual linkages in the chromosomes, although in a number of experiments involving such transformations the introduced marker genes do appear to be stably integrated along with the vector. This probably means that regulatory sequences ligated into a plasmid adjacent to or even remote from the gene will remain in place during integration.

If genes with and without appropriate DNA methylation patterns can be introduced into oocytes, eggs or early embryos, these and similar experimental systems provide the direct approach that has long been needed to answer some of the long-standing questions concerning activation and inactivation of genes during differentiation.

## E. Genome Modifications Which Can Be Associated with Differentiation

### 1. The Insect Type of Differentiation

Whether one examines the morphology, the physiology or the biochemistry of cells, there are many changes which have been studied in great detail and there is much to be learned by studying the pattern and mechanisms of these changes. However, we will confine this discussion to the changes in the genes and other regulatory sequences in the DNA. These are genotypic changes or modifications that can persist through many cell divisions and may not be readily reversible. Differentiation will be considered to involve a modification of the genotype if one or more base pairs have been changed in a way that can be inherited by the daughter cells or a rearrangement of a nucleotide sequence within or adjacent to gene in question has occurred so that the expression is affected. By setting such limits we will confine our discussion to a few cases where information is available. The terminology of those who have studied development and differentiation will be used where applicable. It is not my intention to invent new terminology, but to use a limited amount of the classical terminology when it will simplify the presentation. Modifications not presently known may prove to be important in differentiation and the mechanisms that have been discovered in recent years may prove to be less important than we now think, but we must work with those available.

The terminology and concepts adopted by HADORN (1963, 1965) and GEHRING (1968) are based on extensive experimental data and are very useful in thinking about the problems. The development of cell type differences which result from the division of one cell (usually a zygote or stem cell) is designated cell differentiation. Differentiation is initiated by a process of determination, which programs cells for their future developmental pathways, and may or may not involve subsequent determination steps. The cell may proceed through many cell divisions after determination before differentiation is completed or before a recognizable phenotypic change can be detected. These concepts are based on studies of imaginal disks of *Drosophila* larvae. The disks are bundles of undifferentiated cells which appear as invaginations or thickenings of the epidermis in the later embryonic stages. The disks may also include mesodermal cells as well as those of the ectoderm. During larval stages the disks grow by cell division and acquire a definite shape characteristic for each kind of disk. The disks are composed of layers of epithelial cells which fold during growth and enclose a narrow lumen. The outer surface is covered by a basement membrane and all of the cells are similar and have a relatively large nucleus and little cytoplasm. The cells are basophilic and only about 5 μm in diameter.

When the larvae undergo metamorphosis the disks give rise to specific structures of the adult fly. For example, there are three disks which form the head. The adult thorax and its appendages arise from three pairs of leg disks and three pairs of dorsal thoracic disks. The integument of the abdomen is derived from small groups of imaginal cells or histoblasts and the genital apparatus is formed from a single unpaired disk. Separate clusters of imaginal cells have also been identified for internal organs such as the gut, salivary glands and gonads. HADORN found that these disks could be transplanted and cultured indefinitely in the abdomen of adult flies. The disks continued to grow and could be removed before the end of the fly's life span, 2–4 weeks, and cut into fragments. Those fragments transplanted back into larvae, which were allowed to undergo normal metamorphosis, developed into organs characteristic of the particular disk. Differentiation occurred according to the determination or pattern of change which had been programmed during the formation of the disk. The program of development was maintained in disks which had been cultured by transfers to many adult flies over a period of years.

Further experiments proved that the disks had an organization of cells which produced the different parts of an organ and detailed maps were prepared for some of these. The fragmentation of the disks required for their serial culture proved the mosaic nature of the disks. For example, detailed maps showed that a small sense organ in the third antennal segment, the sacculus, is derived from the anterior half of the antennal disk. In another instance in the male foreleg disk, a region was mapped which determines a single bristle (NOTHINGER and SCHUBIGER 1966). The results indicate that determination is a stepwise process, which progressively restricts the potential of cells within the disks.

To study the state of determination of individual cells, disks were dissociated with trypsin and mechanical manipulations. These dissociated cells were allowed to reaggregate and tested in hosts during metamorphosis. The reaggregated cells either formed organs of one genotype or genetic mosaics of both genotypes used in the mixture. Color mutants of the body such as *yellow* and *ebony* which produce brighter yellow or darker pigmented areas than the wild type were used as markers. The *singed* mutant which has an altered bristle could also be used to follow the differentiation of small regions. Another mutant with multiple wing hairs, in which single cell hairs (trichomes) on a bristle shaft formed from a single cell is replaced by groups of trichomes, provides a way to follow the differentiation of single cells.

In experiments with dissociated cells, mosiac structures composed of cells with different genotypes will usually form only when isotypic cells reaggregate. For example, anal plate cells will form mosaics with other anal plate cells of a different genotype or sex, but fail to associate with clasper cells. The conclusion is that anal plate cells and clasper cells even though both are derived from genital disks already differ from each other in some specific, non-reversible way. Heterotypic cells separate from other heterotypic cells, but isotypic cells appear to migrate and aggregate by selective adhesion. In rare cases separate cells or small groups of cells of one isotype will be isolated into "faulty" mosaics. For example, a single genital disk cell, or small groups of cells, is trapped in a large area of wing disk cells. When this happens, the single cell or small group differentiates autonomously. A specific genital bristle may form. Since the surrounding blastema cells do not exert a determinative influence even upon a single isolated cell, the determination of its specific quality must be due to something the cell carries rather than a contribution of the larger group in which it finds itself.

## 2. Is the Determined State Reversible?

In long-term cultures of the disks, occasional cells reverted to another allotype. For example, cultured genital disks would rarely produce cells which proliferated and later gave rise to antennal or leg structures. The process leading to such variations was termed transdetermination. By means of marker genes, *yellow, ebony,* and *multiple wing hairs,* the transdetermined cells were shown to be derived from the transplanted disks and not form invading cells. The transdetermination events produced cells as stable as those in the original state and the frequency of the event is rare but more frequent than mutations. Clonal analysis was carried out either by inducing mutations in single disk cells by X-rays or by the use of genetic stocks where single marked cells could arise by somatic crossing over. When such disks were cultured and allowed to undergo many cell divisions before differentiation, the evidence indicated that the determined state was a property of individual cells and was transmitted as any other genetic trait. The carrier or the nature of the change to a determined state was unknown, but the experimental evidence indicated that changes were probably associated with cell division, *i.e.,* cell division and probably DNA replication is a prerequisite for determination as well as transdetermination.

## 3. The Vertebrate Type of Differentiation

In the vertebrates determination of cells comparable to that in insects does not appear to occur. Certainly there are no packets of cells comparable to the imaginal disks. The segregation of cells of the germline and soma appears to be the earliest morphogenetic example of a change from totipotency to pluripotency (limitation on the developmental or differentiation potential of cells). The best evidence comes from studies of changes which inactivate one of the X chromosomes of the females in mammals, a topic which will be covered in more detail later. The evidence from the manipulations of teratocarcinoma cells and their totipotency when transplanted into mouse blastocysts (already described) appears to exclude any early determination for, at least, a certain population of cells. Because of the way cell lines have been isolated from teratocarcinomas, there is the potential for a great deal of cell selection and even events comparable to transdetermination may occur and be missed before cells are isolated and tested for totipotency. One may also cite transplantation studies with *Xenopus* as evidence against a determined state in vertebrates. However, in these experiments there is considerable selection before the critical test is possible. The cells selected for the transplants are relatively undifferentiated cells which are mitotically active and the results were achieved by recycling the cells through the blastula stage (GURDON 1962a, b, 1963).

However, vertebrates do have cells in what might be termed a determined state. These are transformed cells which can be induced to produce a certain product by induction, *i.e.,* manipulation of the cellular environment. One case that has been studied extensively is the transformed erythroleukemic cell isolated by FRIEND *et al.* (1966). It can be induced by many treatments, most or all of which block cell division directly or indirectly, to produce considerable quantities of hemoglobin even though it does not differentiate into an erythrocyte (CHRISTMAN *et al.* 1977, 1980). Others are the lymphoma cells which produce monomer IgM as membrane receptors, but do not secrete pentamer IgM unless fused with a plasma cell which produces IgC and J chains (YAGI and KOSHLAND 1981). All of these are transformed cells which appear to be interrupted in differentiation and are "frozen" in some intermediate state. A few of them can be induced to differentiate to a limited extent, but it is difficult to find how far this differentiation can proceed and whether the determined state is comparable to that in the imaginal disk cells of the insect, for example. If these cells, or ones in a comparable state of differentiation, could be placed in the appropriate cellular environment such as the blastocysts used by MINTZ (1978), the differentiation potential might be revealed. However, if the cells failed to participate in development, as is likely, no information would be gained.

## 4. Modifications of the Genome

Modifications during differentiation can be classified as phenotypic or genotypic. If we use the concepts developed from the insect model the genotypic changes might be the events referred to as determination. The phenotypic changes are the actual events of differentiation as the predetermined state is expressed under the influence of a new hormonal environment or some change in the cellular environment that initiates differentiation. The only types of changes which will be discussed are those which alter a base pair in the DNA or induce a rearrangement of the nucleotide sequence within or adjacent to the gene or genes involved. If the change is only phenotypic the expression of the gene involved has been changed, but its structure remains intact. This would include interactions with proteins and cofactors which might alter the secondary or tertiary structure of the DNA without affecting its primary structure. Such changes, or at least some of them, could be long lasting and could possibly persist through replication cycles (MONOD and JACOB 1961). Since specific models are not available, except possibly those which sensitize a region 5′ to the coding region of some functional genes to DNase I (WEISBROD 1982), we will consider only changes in the primary sequence. Such changes, of course, may in turn result in changes in the secondary and tertiary structure of DNA or chromatin.

Two types of modification of the primary structure of DNA have been studied rather extensively. The first involves the mobile genetic elements whose existence had long been suspected on genetic grounds, but only in recent years have been demonstrated and attracted the attention of not only molecular biologists, but a much wider audience. The second is DNA methylation, which has many possible functions, but in the last few years has emerged as an interesting candidate for a controlling factor in differentiation of vertebrates, at least.

### 5. Mobile Genetic Elements

The meaning of this term, mobile genetic elements, is best given by examples rather than by a definition. There are possibly many classes of these sequences in DNA and what was once thought to be a relatively static structure, the genome, is now visualized as subject to considerable rearrangement over evolutionary time and possibly over the time of the life of single individuals, which would be the case if the elements are factors in differentiation. The mobile genetic elements are not the primary object of this monograph, but an evaluation of the role of DNA methylation requires an evaluation of the principal competing mechanism which involves mobile genetic elements. Competition may be the wrong term, for the two mechanism are not competitive in any sense except in the minds of the scientist. Both mechanism might operate simultaneously at different sites in the genome and it is not inconceivable that the two systems could be functionally connected.

Beginning in the 1940's McCLINTOCK (1951) reported changes in the expression of genes in *Zea mays* (corn) which indicated that chromosomal elements, presumably DNA, moved to the vicinity of a gene by some type of translocation mechanism involving chromosome breakage. The movements were relatively rare events but higher in frequency than regular mutations. The changes that could be detected resembled mutation in that the affected gene was no longer expressed in the tissue derived from the cell where the original event occurred. The gene might be suppressed for many cell generations and even passed on through the seed, but the structure of the gene was shown to be intact. At some subsequent cell division, perhaps in succeeding generations of plants, the gene might be switched on again. This appeared to involve the translocation of the movable genetic element to another site in the genome. The uncovering of the gene was the event that could be followed experimentally. The initial event that resulted in suppression or mutation of the gene was a spontaneous one that was unpredictable and rare. However, once the mutants, which involved the change from a functional to a non-functional state, were recognized the reversal could be detected in cells of various tissues; frequently the endosperm was the choice for visualizing the reverse mutation.

### a) Examples of Mobile Genetic Elements in Corn

McCLINTOCK (1950, 1951) was studying stocks of corn which had undergone the chromosome type of breakage-fusion-bridge cycle during their early development. This cycle is initiated when two chromosomes with ends that have undergone breakage are delivered to a zygote. The broken chromosome in this instance was chromosome 9 which was delivered to the gametes when the sporocytes were heterozygous for an inverted duplication of the short arm of chromosome 9. When chromosome 9 with a complete set of genes and an open break was delivered to the microspore or megaspore it would have nothing to join until chromosome replication occurred. Then it would join its sister chromatid and form a chromatid bridge at the subsequent mitosis. It would break and repeat the cycle at the next mitosis. In the haploid cell of the gametophyte, with one broken chromosome, this cycle called the chromatid type breakage-fusion-bridge cycle would continue until the formation of either the sperm or the egg nucleus. Since the break at mitosis would not necessarily be exactly at the point of fusion, there would be deletions and duplications usually toward the ends of the short arm of chromosome 9. If a broken chromosome from a microsporocyte were delivered in each sperm nucleus of a pollen tube entering an embryo sac at fertilization, the products would each have one broken chromosome. Both the zygote and the triploid nucleus that forms endosperm would have a single broken chromosome if the female parent were normal. The chromatid type of breakage-fusion-bridge cycle would continue after fertilization in the development of the endosperm but in the embryo the chromosome end would heal and fusion and breakage would cease. If the short arm of chromosome 9 involved in the breakage were properly marked with genetic markers the deletions could be followed by the pattern of colored sectors or other traits visible in the endosperm. The markers used by McCLINTOCK were mutants which had been mapped along the short arm from the end toward the centromere in the following order: *yg* (yellow-green chlorophyll in the young seedlings), *c* (colorless or lacking the red anthocyanin); *I* (a dominant allele at the same locus also inhibited anthocyanin), *sh* (shrunken endosperm, a reduction in the starch content), *bz* (a change from the red or purple endosperm or plant color to a bronze-shade) and *wx* (a change in the starch so that iodine stains the endosperm reddish brown instead of the standard blue).

When a broken chromosome 9 is delivered by fertilization from both sperm and egg nuclei, the two broken ends fuse before chromosome replication and the chromosome-type of break-fusion-bridge cycle continues. Many of the terminal deletions result in cell death so that the plant grows very poorly with many shoots instead of the single stalk typical of most corn. However, for reasons unknown the ends heal in an occasional cell and the breakage cycle stops. Shoots derived from these cells will produce ears and tassels and it was microsporocytes of such plants that were examined by McCLINTOCK.

In the endosperm both the chromatid and the chromosome-type of fusion-breakage-bridge cycle continue, but since endosperm has 3 sets of chromosomes and probably requires fewer genes for survival than embryos, development proceeds in a fairly normal fashion. By the use of proper genotypes in crosses, the pattern of deletions can be followed. For example, if the female parent is homozygous for *c* and the male carries the *C* gene, the endosperm will be red with colorless spots where deletions have shortened the chromosome enough to include *C* of that one chromosome 9. One advantage of using corn is that the genotype of the embryo and endosperm are identical except that the endosperm has two sets of chromosome from the female parent. The two sperm nuclei from a pollen tube which participate in the double fertilization are genetically identical as are the nuclei from the female parent with which these fuse. Therefore, if one finds an interesting chromosome in the endosperm as indicated by its effect on the phenotype, the embryo will have received the same chromosome. By planting the seed and growing the plant to maturity, the chromosomes can be examined at the pachytene stage of meiosis and seed can be collected from the same plant to maintain the stock.

In the progeny of one of the plants that had undergone the chromosome type of breakage-fusion-bridge cycle, McCLINTOCK found an unusual number of breaks at one locus on the short arm of chromosome 9 at the pachytene stage of meiosis (Fig. 5 *A*). Breaks were localized at a specific site in the short arm just proximal to the *Wx* locus. The symbol *Ds* (dissociation)

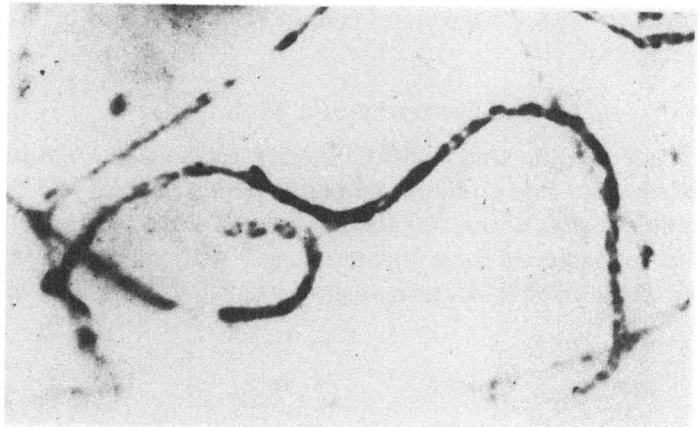

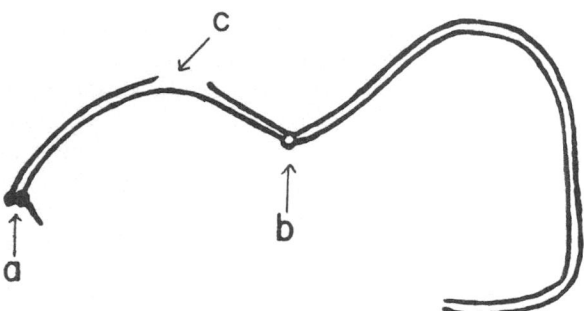

Fig. 5 *A*. Chromosome 9 at the pachytene stage of meiosis in corn (*Zea mays*); photograph (upper) and diagram of the same chromosome (lower). Arrows indicate a knob and a small projection on one homolog (*a*), centromere (*b*) and break (*c*) at the Ds locus. (From McClintock, 1951; original photograph provided by Dr. Barbara McClintock)

was assigned to this locus. Many kinds of aberrations involving breaks at this locus were seen in different cells. However, the frequency of breakage events varied among the plants descended from the original ones.

In attempts to map *Ds* using standard procedures, she found that *Ds* was transposed from one position to another on the same or different chromosomes of the complement. The standard position was designated as the one in which it was first detected, *i.e.*, just proximal to *Wx*. In the process of making the crosses to map *Ds*, another heritable factor necessary to its operation was found. It was a dominant factor designated activator (*Ac*). *Ac* was inherited in a Mendelian fashion as if composed of a single genetic unit. At all sites *Ds* required *Ac* for mutability. When *Ac* was segregated, *Ds* no longer produced beaks, but when brought together again by crossing, the breakage resumed.

## b) The Discovery of a Mutable Gene

In the process of the studies McCLINTOCK recognized an important and significant mutant. In a cross of one plant (pollen parent) to 12 genetically similar female plants about 4,000 corn kernals were examined. The male parent was homozygous for a chromosome 9 with genetic markers *Yg, C, Sh, wx* and *Ds* (standard location as shown in Fig. 5 *B*) and an unlinked *Ac*.

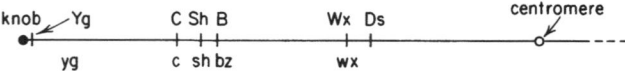

Fig. 5 *B*. Linkage map of the short arm of chromosome 9 in corn. The symbols represent genes that result in the following traits in plant or endosperm aleurone layer. Alleles *Yg*, normal chlorophyll color and *yg*, yellow-green color in early development of the seedling. Alleles *I*, no color (dominant or epistatic to all aleurone color genes), *C*, full color when homozygous or heterozygous, and *c* no color when homozygous. Alleles *Sh*, full kernel and *sh*, shrunken kernel. Alleles *Bz* which changes red aleurone to bronze, and *bz* with no change. Alleles *Wx*, starchy endosperm, and *wx*, waxy endosperm (a defect in carbohydrate metabolism). *Ds* is the dissociation (breakage) factor that can cause the loss of all genes distal to its position. (Redrawn from McCLINTOCK, 1951)

The female carried the recessives, *yg, c, sh, wx* and no *Ac*. Approximately one half of the kernels were *C Sh wx* as expected and without variegation (Fig. 5 *C*). The other half which had inherited *Ac* showed some variegation due to breakage and loss of the short arm of chromosome 9. Sectors with the *c sh* phenotype were found. The white areas in a red background were easily detected, but one unusual kernel was seen and recognized for its value. Instead of a colorless area in a red background, it had a colorless background with red dots (Fig. 3 *C*). The seed was planted and it grew into a plant which showed that the *C* on the chromosome from the male parent had mutated to *c* and mutations back to *C* occurred only in the presence of *Ac* in the genotype (Fig. 5 *C* and 5 *D*). *Ds*-type activity was occurring in this chromosome with the new *c* mutant and attempts to map it showed that it could not be separated from *c*; it was very closely linked. *Ds*-activity had moved from the standard location and suppressed *C* to make it an unstable gene which was designated $c^{m-1}$.

Here was an explanation for spontaneously mutable loci in plants and an indication that a movable element induced the change. When $c^{m-1}$ mutated to *C* the breakage at that locus ceased; *Ds* had moved. Other mutants were later recognized and studied at the various loci on chromosome 9, as well as

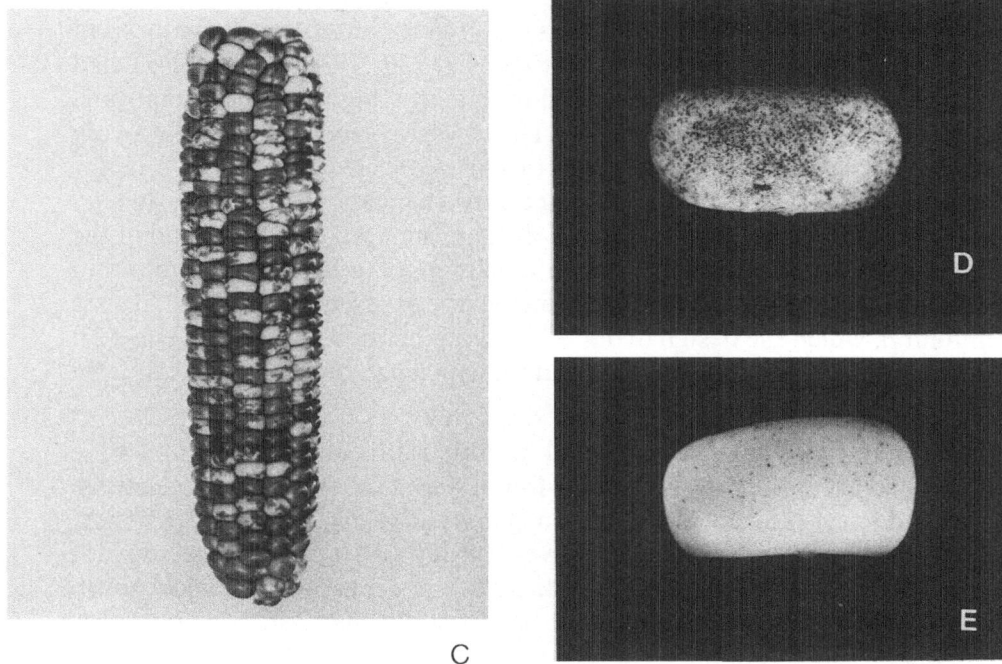

C

Fig. 5 *C*. An ear of corn segregating for the Ac factor (variegation and fully colored, 1:1). The plant with *c* (lack of color) in each chromosome 9 and no *Ac* factor was pollinated with pollen which carried *C* and *Ds* (standard location) and one *Ac* factor. Most kernals with endosperm derived from a nucleus with the Ac factor are variegated due to occasional loss of a segment of chromosome 9 containing the *C* gene at divisions in development of the endosperm. (From McCLINTOCK, 1951; original photograph provided by Dr. BARBARA McCLINTOCK)

Fig. 5 *D* and *E*. Two kernels of corn showing variegation when there are two (*D*) or three (*E*) doses of the *Ac* factor in the presence of *Ds*. The kernels were on a ear from a plant carrying *C* and *bz* in each chromosome 9. The grain would be colored except that the pollen carried I, *Bz* and *Ds* at it standard location (Fig. 5 *B*). The I gene prevents any color development except in areas where it is lost from the cells due to breakage at *Ds*. When there are three doses of *Ac* the breakage occurs late in development and only a few small colored dots appear in the endosperm. This pattern is very similar to the action of *Ds* when it is sitting next to *C,* for example ($c^{m-1}$) or to dotted when *Dt* acts on an unknown factor similar to *Ds* next to *A* (another gene necessary for color development in the aleurone). In both cases the white background develops dots of color when the inhibiting factor (mobile genetic unit) moves away from *A* or *C* by translocation, a rare but variable event depending on the doses of *Ac* in the presence of $c^{m-1}$. (From McCLINTOCK, 1951; original photograph provided by Dr. BARBARA McCLINTOCK)

other chromosomes in the complement. An interesting one involved the gene $A$, which along with $C$ and $R$, is required for anthocyanin synthesis in *Zea mays*. Some years before, RHOADES (1936, 1938, 1941, 1945) had discovered a gene which caused $a$ to mutate at a higher than normal rate, *i.e.*, converted $a$ to a mutable locus. The recessive gene $a$ is normally a very stable locus, but in certain stocks of corn collected in Mexico a factor designated Dotted ($Dt$) when crossed into the standard colorless (*white*) stocks made $a$ mutate to the dominant $A$, *i.e.*, on a white background of the kernels, red dots appeared. These consisted of cells with anthocyanin pigment. Strangely $Dt$ had no effect on the rate of reversal of McCLINTOCK's mutant $a$, which she designated $a^{m-1}$, and furthermore $Ac$ had no effect on the mutability of the standard $a^1$ mutant which had been made unstable by $Dt$.

Mutable genes have been studied in both plants and animals and have continued to be of interest wherever found. For a more complete review see the Cold Spring Harbor Symposia on Quantitative Biology, Vol. XLV, part 2, Movable Genetic Elements. However, McCLINTOCK's interest was not directed to a mutation mechanism, but to a differentiation mechanism which, she believed, had become aberrant in the system under consideration. She emphasized the variations in the timing of the interactions of $Ds$ and $Ac$. One variation was related to the dose of $Ac$ which could be varied from one to three in the endosperm (Fig. 5 $D$ and $E$). In addition there were different states of the two inacting components of the system ($Ac$ and $Ds$), which appeared to affect the timing of the $Ds$ activity and the expression of the gene involved. It is the possibility that such two-component systems have a role in differentiation, which has kept interest alive and made it seem worth spending so much space on the topic in this monograph on DNA methylation.

The movable elements in higher plants have recently been identified (BURR and BURR 1982) but the nature of the DNA sequences involved is still uncertain. Several of the systems are being investigated vigorously in a number of laboratories and before this monograph can be published, it is likely that much more will be known about the movable elements in maize. However, in bacteria with simpler genomes where mutants can be manipulated and generations are measured in hours rather than months, the insertion elements and transposons were demonstrated by genetic experiments and many of the elements have now been isolated and sequenced. Model systems similar to those indicated for higher plants are known and the way is open for much more complete analysis. PEACOCK and colleagues have isolated and sequenced a 402 bp sequence which has moved into a gene that codes for the enzyme alcohol dehydrogenase in maize. The insert has 11 bp inverted repeats and 8 bp direct repeats at the ends beyond the inverted ones. Since it moved into the gene under the control of

McCLINTOCK's *Ac* gene, it should be a *Ds* element (see Science 219, 829—830). On the other hand NINA FEDEROFF and coworkers found a 2,500 bp *Ds*-like element which moved into the waxy (*Wx*) locus that codes for the enzyme amylase synthetase. The revertants did not remove the element precisely since the protein product of the revertant gene was larger than the original enzyme.

The insertions reported by BURR and BURR (1982) were much larger, about 20,000 bp. A follow-up study of one of the insertions by FEDEROFF and also by STALINGER in Cologne also indicated a large element, but in this case the mutant gene had a partial duplication of the gene as well as a newly introduced sequence. This would be another *Ds* element since it moved into the shrunken locus under the influence of *Ac* and in one revertant the element is gone but the duplicated segment remains. At this time the size and exact nature of *Ds* is still uncertain.

# II. DNA Methylation and Transposable Genetic Elements

## A. Discovery of Enzymatic Methylation of DNA

The process of DNA methylation has excited the imagination of many investigators since GOLD *et al.* (1963 a, b) demonstrated an enzyme which transferred methyl groups from S-adenosylmethionine (SAM) to DNA. The feasibility of this process had been indicated by the discovery that a similar transfer was catalyzed in vitro by soluble enzymes from *Escherichia coli* and a variety of cells from higher animals and plants (BOREK *et al.* 1964, and BOREK 1963). The appropriate substrate for the in vitro reaction was found by accident when *E. coli* K 12 W 6 was deprived of an essential nutrient, methionine. Unlike most strains of bacteria which stop growth and RNA synthesis when starved of an essential amino acid, this one continues to produce RNA for a time even though it can no longer methylate the RNA because the precursor SAM requires methionine for its regeneration. Such strains have been called "relaxed" since there is a loss of the normal control of RNA transcription. The RNA isolated from the relaxed strain after methionine starvation was found to readily accept methyl groups when incubated with a soluble cell extract and SAM, that had been labeled in the methyl group derived from the $^{14}$C-methionine residue. The small RNA made by *E. coli* K 12 W 6 in vivo was isolated from the reaction mixture and analyzed for radioactive ribothymidylate. The finding of this methylated uridylate labeled with $^{14}$C indicated that the RNA was methylated at the polymer level since no RNA synthesis was occurring in the in vitro reaction. Later, other bases were found to be labeled and the enzymes catalyzing the reaction were shown to be present in cells from a number of species of animals and plants. In animals, tissues from various organs were also shown to have the methyl transferases. By demonstrating that an RNA, which was completely methylated in one species could, nevertheless, accept methyl groups when reacted with a crude enzyme preparation from a different species, they were able to show that the methylases had species specificity.

After GOLD *et al.* (1963 a, b) reported the methylation of DNA by a similar activity specific for DNA, BOREK speculated about the biological role of such enzymes. No suggestions of lasting significance were mentioned, but he thought that aberrant methylations, which had recently been shown to be a feature of certain alkylating carcinogens, might be a factor in causing cancer. GOLD and HURWITZ (1963) suggested three possible mechanisms in which methylation might have a role: 1. replication of DNA; 2. transcription of messenger RNAs in which the methylated bases might act as sites of initiation or termination of the transcript; and 3. a nuclease recognition system in the prophage immunity system of *E. coli* which was under investigation by ARBER and DUSSOIX (1962).

## B. The First Demonstrated Role of DNA Methylation: Restriction-Modification Systems in Bacteria

HOST controlled modifications of phage were discovered by LURIA and HUMAN (1952) when they showed that a phage grown in one strain of bacteria could reinfect cells of that strain, but if plated on a related strain the infectivity could be very low. Other studies soon revealed that such host range restrictions were a common phenomenon. For example, phage lambda plates with high efficiency on *E. coli* K 12, but if the same strain becomes lysogenized by phage P 1, the new strain now called K (P 1), is very resistant to phages grown on the original K 12. The efficiency of infection is only about one plaque per $10^4$ phage particles. However, phage from those few plaques which survive will now plate efficiently on K (P 1) as well as on the original K 12 strain. One growth cycle on K (P 1) is sufficient to acquire immunity, but after a subsequent growth cycle on K 12 the immunity is lost. The modification does not survive the replication cycles in a non-modifying host. A clue to the type of modification was obtained by ARBER (1965) and ARBER and LINN (1969) when they found that phage lambda grown in K 12 host cells was equally effective in plating on either K 12 or *E. coli* C which is not restrictive. However, if the phages were produced on a mutant of K 12 which required methionine an exception in restriction could be induced. If ethionine was substituted for the amino acid methionine the early maturing particles, produced by premature lysis after restoring the methionine to allow some additional growth, were unable to grow on K 12, but could grow on *E. coli* C. It was by then known that methylation required S-adenosylmethione (SAM) as a methyl donor for methylation and ethionine could not substitute for methionine in these reactions. They were able to proceed with these clues and finally to isolate the modifying activity and the restriction endonuclease from K 12.

These enzymes represent what are now called Type I restriction enzymes. The molecules are very large and contain at least three subunits. One has nuclease activity, another methylase activity and the third is a site recognition factor. The enzymes have a strict requirement for both ATP and S-adenosylmethionine in addition to $Mg^{++}$. There are identifiable recognition sites but the enzymes cut the DNA at a point more than 1,000 bp away from the recognition site. The cleavage in vitro is followed by a massive hydrolysis of ATP of unknown significance, but without additional cleavage of DNA. The recognition site for the Eco B Type I restriction enzymes has been identified by RAVETCH *et al.* (1978) as shown below:

5′-TGANNNNNNNNTGCT-3′

The recognition sequence is TGA separated by eight nucleotides from TGCT.

The Type II restriction enzymes are the more familiar ones used in recombinant DNA studies. The recognition sites usually consist of 4 or 6 nucleotides with palindromic characteristics, *i.e.,* the sites usually have twofold rotational symmetry and the cleavage site is precise for each enzyme and within or adjacent to the recognition site. At least 398 are now known that recognize 97 different sequences (ROBERTS 1983). They were first characterized by SMITH and WILCOX (1970) and KELLY and SMITH (1970). The two activities, endonuclease and methylating activity reside in separate molecules. We will discuss some of these restriction enzymes at greater length in subsequent chapters.

A third group which may be considered intermediate between Type I and Type II, has been recognized, but is now considered a distinct Type III. Examples are the two *E. coli* enzymes Eco P 1 and Eco P 15. Eco P 1 is coded by phage P 1 and the other by a plasmid, P 15. The recognition site for Eco P 1 is AGACC. These enzymes molecules have both endonuclease and methylating activities, but the activities can be controlled by the available co-factors. In the presence of SAM the methylase is active. When ATP and $Mg^{++}$ are present, the endonuclease is active, but when all three factors are present the two activities, methylation and cleavage, are competitive. A third enzyme in this group, which has been recently characterized is from *Haemophilus influenza* Rf and is named Hinf III. It recognizes the non-symmetrical site 5′-CGAAT-3′ and methylates one chain of the DNA at the middle A. It cleaves 25–27 bp to the 3′ side of the methylatable base and leaves the DNA with 5′-protrusions (BACHI *et al.* 1979, PIEKAROWICZ *et al.* 1980).

The roles that methylation plays are revealed in some instances, but there remain many questions. Certainly the Type I enzymes typical of *E. coli* K and *E. coli* B function in restricting the survival of DNA entering by phage infection or transformation. The Type III enzymes can serve a similar role, but the Type II enzymes are more puzzling. Two of the most extensive systems are those revealed by two mutations which delete the respective enzyme. The *dam*⁻ mutants lack the enzyme to methylate adenine in the symmetric tetranucleotide d(pGATC) and a similar molecule, deleted in dcm⁻ mutants, methylates cytosine in the Eco RII site, CC(A/T)GG. The dam⁻ mutants appear to be more susceptible to certain mutagens, but otherwise grow quite normally (GLICKMAN *et al.* 1978). RADMAN *et al.* (1980) and GLICKMAN and RADMAN (1980) have provided evidence that this type of methylation may play a role in determining which chain is repaired when there is a base pair mismatch. This role will be considered in detail later.

## C. Hypotheses for Roles of Methylation in Eukaryotes

Eukaryotes are a more heterogeneous group than prokaryotes and methylation may have evolved to serve different roles than in bacteria. Even in bacteria methylation is unlikely to have a single role. One role appears to be related to restriction-modification systems of Type I and III. The protection of sites from cleavage by Type II restriction endonucleases is another important role. Methylation of DNA is also believed to play a role in recombination and there is evidence that methyladenine allows strand recognition in mismatch repair as mentioned above.

Restriction-modification systems in eukaryotes similar to those in bacteria appear to be rare. SAGER and KITCHIN (1975) presented a case for the operation of such a system in the destruction of the chloroplast genome of one parent (the minus mating type) in the zygote of *Chlamydomonas*. Subsequent evidence supports the hypothesis (SAGER *et al.* 1981) but the matter is still under active investigation and there seems to be serious doubts that the present evidence supports such a role (BOLEN *et al.* 1982, DYER 1982). SAGER and KITCHEN also speculated that such a mechanism could be involved in the uniparental inheritance of not only chloroplasts but mitochondria. Since these organelles are thought to have descended from primitive prokaryotes, the persistence of a restriction enzyme system might have occurred. However, such unrelated phenomena as chromosome elimination in interspecific somatic cell hybrids, nuclear destruction that follows plasmodial fusion in *Physarum*, haploidization in barley, elimination of B chromosomes, and heterochromatization and chromosome elimination in insects, which they suggested is less likely to be explained by restriction-modification systems similar to eukaryotes.

Now that we know more about the patterns of DNA methylation in higher eukaryotes than was known in 1975, we can eliminate restriction-modification as a functional mechanism in most higher organisms. Even then it was known that most of the 5mC in eukaryotes is in the dinucleotide, CpG, (DOSKOCIL and SORM 1962, and GRIPPO *et al.* 1968), and this dinucleotide may have a variety of bases as nearest neighbors. All of these CpG sequences are incompletely methylated, unlike the DNA in bacteria. If there were restriction enzymes, the whole genome would probably be continuously cleaved at such unmethylated sites.

The first extensive and well documented hypothesis for a role of methylation in higher eukaryotes has been presented by SCARANO and associates (SCARANO *et al.* 1965 and 1967, GRIPPO *et al.* 1968, GRIPPO *et al.* 1970, TOCI *et al.* 1972, and SCARANO *et al.* 1977). They studied the metabolism of thymidylate in sea urchin, but the hypothesis was assumed to apply to all higher organisms. Thymidylate is synthesized in all cells by a process in which the 5-methyl group is added to deoxyuridylate by the enzyme thymidylate synthetase. The methyl donor is tetrahydrofolate. It also contributes methyl groups in the synthesis of the purines and in the metabolism of some other small molecules. On the other hand the methyl donor for macromolecules such as proteins, lipids and nucleic acids is S-adenosylmethionine (SAM) and the variety of enzymes that catalyze the transfer are called methyl transferases. DNA methylases are one of these classes. SAM contributes the methyl group to DNA cytosine and adenine when the addition occurs at the polymer level. They obtained evidence that thymine derives its methyl group from two sources. Most of it comes from the one carbon pool by way of tetrahydrofolate. However, a small amount, minor thymine, appeared to be contributed by SAM. They proposed that this fraction might come from the methylation of cytosine in DNA and its later deamination to produce thymine. After replication, the C:G base pair would be converted to a T:A pair in one chromatid. This would be a non-reversible change or directed mutation presumably catalyzed by a specific deaminase. This activity was never demonstrated in vitro.

One may summarize all of the experiments from SCARANO's lab by saying that they have rather convincing evidence that thymine in DNA gets its methyl groups from two sources: 1. the one carbon pool via tetrahydrofolate, *i.e.,* from the synthesis of one precursor for DNA, thymidine triphosphate and 2. a minor fraction, perhaps no more than one percent via SAM. Since the minor thymine is a small fraction of the total it might be dismissed as an impurity or deamination product in hydrolysis of DNA to bases. However, all of these possibilities have been considered and controls run (GRIPPO *et al.* 1970). From my study of the data in four or five of the research papers I think there may be methylation of a base at the polymer level to give some small fraction of thymine that derives its methyl group at the 5-position from methionine via SAM. If we admit the possibility of this minor thymine, there are at least two possible origins: 1. the first, and the one favored by SCARANO, is methylation of cytosine and then enzymatic deamination of the 5-methylcytosine to produce thymine; finally following replication, an A:T base pair would appear where G:5mC existed after methylation and before deamination, and 2. the methylation of uracil in DNA to produce thymine directly. SCARANO and his colleagues have continuously emphasized the first because they were looking for a programmed change that would make the gene respond to a different control mechanism—a determined cell which would differentiate at some later time. Attempts to demonstrate a deaminase that would catalyze the reaction in DNA have so far been unsuccessful, but perhaps no one has produced the appropriate substrate for the test. The DNA of the phage XP-12 of *Xanthomonas oryzae* would be an ideal substrate to search for such an enzyme. All of the cytosines are substituted with 5-methylcytosines (EHRLICH *et al.* 1975). However, the search is likely to be fruitless. Such an enzyme would deaminate DNA at many sites since there seems to be limited specifity in the sequences around CpG sites which are methylated.

The second possibility, the methylation of uracil in DNA might also seem to be an unlikely mechanism for obtaining the minor thymine. However, deoxyuridylate is incorporated into DNA in detectable amounts when the concentration is abnormally high in cells of *E. coli* and probably eukaryotic cells. The uracil can be removed by uracil-DNA glycosylase (SEKIGUCHI *et al.* 1976) and the apyrimidinic site replaced by the "cut and patch" type of repair, which also serves to replace the uracil with a cytosine containing nucleotide if there is a mismatch due to spontaneous deamination of cytosine. If the uracil resulted from incorporation of the nucleotide from deoxyuridine triphosphate, it could also be removed by the same mechanism. However, in principle, there is no reason why some cells may not have evolved a methylase for converting uracil to thymine in the polymer in the absence of a mismatch. It is also conceivable that deoxyuridine triphosphate is regularly incorporated into the primer sequences for DNA replication and is later converted to thymine by methylation rather than excised and replaced.

Whatever the explanation for the minor thymine may be, it now seems unlikely that the conversion of G:C base pairs to A:T by methylation followed by deamination is a mechanism for differentiation of cells. To get a similar effect on the binding of proteins to DNA, it may not be necessary to have the transition of G:C to A:T as proposed by SCARANO. As we have seen earlier in Chapter I, the binding of the *lac* repressor protein to DNA can be changed just as effectively by methylation of the cytosine in a G:C base pair which causes the site that had mutated from the original A:T to act as if the original A:T pair had been restored.

In 1975 four proposals were made for new roles of DNA methylation. In addition to SAGER and KITCHIN's proposal for protection of chloroplast DNA of one parent in sexual reproduction in *Chlamydomonas,* RIGGS (1975) made a rather good case for a role of DNA methylation in the modification and maintenance of the inactive X chromosome in female mammals. We will cover this topic in more detail in a later section. However, it is well known that one of the X chromosomes in cells of embryos is modified after the blastula stage so that it remains heterochromatic in subsequent cell cycles. It no longer expresses the mapped genes, except one on the short arm of the X chromosome in the human genome, for example. Other mammals have some variation of this scheme for an inactive X in cells of the female. The time of replication of the modified X in the cell cycle changes simultaneously with its genetic inactivation. In addition, there is conversion to a heterochromatic state so that one condensed X chromosome in each cell can be recognized in interphase nuclei as the "Barr body".

RIGGS' model for modification was based on the work of MESELSON and YUAN (1968) who were studying the Type I modification in *E. coli* strains K and B. The enzyme is very slow in catalyzing the initial reaction, *i.e.,* methylation of a previously unmethylated DNA, but hemi-methylated DNA (methylated on one chain) is rapidly methylated on the other chain at symmetrical sites. RIGGS proposed that there are primary inactivation centers on the X chromosome that react very slowly in the presence of methylase, usually after many hours. However, once one of the primary inactivation centers is methylated, the result is an activation of the chromosomal site and the synthesis of two proteins, or a bifunctional protein, is induced. One protein changes the methylase so that no other centers are activated and the second protein functions to condense and inactivate any X chromosome with an unmethylated inactivation center. A necessary assumption is that the changed methylase can now methylate hemimethylated DNA, but can not operate at an unmethylated site. Therefore, the methylase becomes a maintenance methylase which can utilize only hemi-methylated DNA as a substrate and thereby maintain the methylated state after replication, but other inactive chromosomes, one or more in any cell, will remain inactive. The details of this model are probably incorrect, because there is now evidence that methylation may be involved in X-chromosome inactivation rather than activation. In nearly all cases so far reported, methylation appears to be correlated with inactive genes rather than serving a role in activation.

The third proposal, by HOLLIDAY and PUGH (1975), was that methylation could be a mechanism for developmental clocks. It is based on the assumption that tandem repeats in DNA could provide a substrate for two types of methylases, a switch enzyme and a clock enzyme. The switch produces a hemimethylated site on the first member of a tandem array of methylatable sites. The clock enzyme would then react with the site to methylate the second chain and extend the methylation into the next repeat. After the next replication the clock enzyme repeats the original catalytic event. It completes the methylation of the hemimethylated DNA and extends the methylation downstream one repeat. These events occur every time replication occurs. Presumably after a certain number of divisions (replication cycles) the methylation activates or inactivates a transcription unit downstream from the initial switch. Since no evidence for such a mechanism has been uncovered, we will not pursue this clever model here.

In 1975 at the EMBO Workshop on Recombination held at Nethybridge,

Scotland, MESELSON proposed a model in which methylation could serve as a marker for correction of base mismatches in newly replicated DNA. This idea is consistent with reports that bacterial mutants (dam⁻), suspected of being deficient im mismatch repair and an adenine methylase, exhibit spontaneous mutator effects (RADMAN et al. 1980). It was some years before the evidence was obtained and presented in detail. The tests could not be made directly on the newly replicated chains, but heteroduplex phage lambda DNA was prepared without methylation of adenine at the dam sites (GATC), with hemimethylated DNA, and with both chains methylated. Transfection of *E. coli* strain C 600 and a dam⁻, dcm⁻ derivative (lacking methylating activity for both GATC sites and CCA/TGG sites) produced results consistent with the predictions of the model. The mismatch was only extensively repaired in the DNA which was hemimethylated and the unmethylated chain was the one excised and replaced. If such a mechanism operated in *E. coli* just back of the replication fork, it could be an effective proof reading mechanism to reduce mutations during replication.

Some eukaryotic DNAs maintain a high level of CpG sites in spite of the fact that the cytosine is methylated and, therefore, is a potential "hot spot" for mutation. It is possible that in eukaryotes one role of methylation may be to distinguish DNA chains for mismatch repair and perhaps also be a controlling factor in some types of recombination. We will come back to this topic later.

Another role for methylation could be to prevent multiple replication of chromosomes or chromosomal regions during one cell cycle (TAYLOR 1977). Although most methylation of DNA occurs immediately following replication, a significant fraction is delayed until S or G 2 (ADAMS 1971, WOODCOCK et al. 1982). TAYLOR (1977) proposed that inhibition of a second replication in one cell cycle might be a function of methylation. If the proteins which initiate replication fail to bind to hemimethylated DNA and a certain relevant part of methylation around origins were delayed until after all of the DNA is replicated through one cycle, *i.e.* until the G 2 stage, we could understand how eukaryotes might have hundreds of thousands of origins for DNA replication which replicate in a variety of sequences and yet do not often replicate two times in one S phase.

The last and perhaps most important proposal for a function of methylation in eukaryotes is that it serves as one system for regulating RNA transcription. As we shall see, most of the evidence available indicates a correlation between undermethylated or unmethylated DNA and the ability of other regulatory mechanisms to initiate transcription. Highly methylated DNA can be replicated and the methylation pattern is maintained. However, some change, usually involving DNA replication, is necessary before certain genes are available to the transcription apparatus of the cell. We will spend much of the remaining space in this monograph collecting and trying to evaluate the,evidence for and against this hypothesis of a major role for methylation in vertebrates and probably some of the distant ancestors of the vertebrates. One large group of animals the arthropods, and particularly the insects, may have evolved other mechanisms for modifying DNA. Certainly the methylated CpG sites appear to be rare or absent in the best studied example, *Drosophila melanogaster*.

Some may contend that a mechanism that is not universal must be of limited significance. That argument can be advanced for all of the proposed functions of methylation, for there are phylogenetic groups, that are exceptional, since DNA methylation of either cytosine or adenine is so rare that it is undetectable by present methods of analysis. However, let us examine the evidence and the implications of enzymatically methylated bases in DNA and then try to come to grips with these questions of great variability in the degree of methylation in different phylogenetic groups of both prokaryotes and eukaryotes.

## D. Transposable Genetic Elements

### 1. Flagellar Phase Variation of Salmonella

We ended our section on MCCLINTOCK's pioneering work with *Zea mays* which indicated movable genetic elements without giving any specific molecular examples. Indeed, we can not yet define these units in maize or any higher plant, but we can look at a few examples in bacteria to see what properties these genetic elements have and how these might explain the genetic behaviour observed not only in maize but in other eukaryotes as well. Their role in explaining mutable genes will be obvious, but to show how these manipulations of the genotype could operate in differentiation still takes considerable imagination.

We will look first at the change in flagellar phenotypes in the bacterium, *Salmonella* (reviewed by IINO and KUTSUKAKE 1980). For years observations had indicated phase changes in flagella within what we know as disphasic strains. These strains possess two non-allelic genes for the structural protein, flagellin, from which flagella are built. The genes have been designated H 1 and H 2 and a cell usually has one or the other but not both genes operating. Yet the flagellar type can change rather frequently in some strains and rarely in others. Genetic studies of *S. typhimurium* C 77 had indicated changes from H 1 to H 2 at the rate of one cell per $10^5$ each cell division and a little lower shift in the other direction, one in $3 \times 10^4$ per cell generation. These frequencies are higher than normal mutation rates at stable genetic loci, and much higher than the rate of phase change of some mutant strains, for example, *S. abortusequi* SL 23 in which the switches in either direction are less than $10^{-7}$ per generation. The genetic studies indicated that the H 2 structural gene was in an operon with a gene coding a repressor of H 1 (rh 1). When H 2 was turned on, the repressor was produced simultaneously. It acted on the operator locus of H 1 and prevented its transcription, but when the H 2 operon was off the repressor coded by the gene designated rh 1 was also absent and H 1 could be expressed (Fig. 6).

The phase change in *Salmonella* has some features in common with the mutable loci im maize, but its operation is also different and the mechanisms could be different, but let us make some comparisons. The frequency of the changes are of the same order of magnitude. The switches of flagellar types is $3 \times 10^{-4}$ to $10^{-5}$ per cell generation for the standard *S. typhimurium* C 77

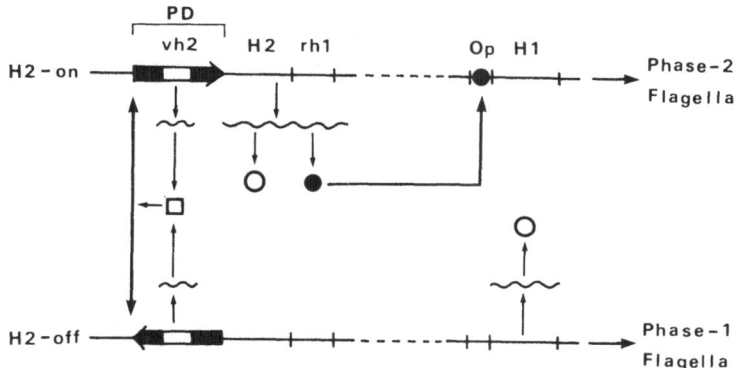

Fig. 6. The genetic map of the *Salmonella* chromosome in the region involved in flagellar phase variation and the products of the genes in the region. (○) flagellin; (●) HI repressor and (□) effector of *PD* inversion. (From IINO and KUTSUKAKE, 1980; copyright by Cold Spring Harbor Laboratory, 1980)

strain. The rates for maize are less precisely known; recall that the rate could vary over a wide range depending on the dose of *Ac* (the activator) and the state of the mutable locus (McCLINTOCK 1951). In any case the rates of change at mutable loci in maize are two or three orders of magnitude greater than the mutation of stable genes which is between $10^{-6}$ and $10^{-7}$ (STADLER 1948). The change in maize is under the control of an activator (*Ac*) locus, which can be anywhere in the complement, while in the flagellar phase change the regulator of the phase change turned out to be linked to the H 2 operon (designated *vh 2* in Fig. 6). The nature of the activator in *Salmonella* did not become clear until the genes were cloned and analyzed genetically and physically, but IINO (1961) had discovered the activator when he isolated a mutant designated *vh 2⁻* in which the phase variation was less than $10^{-7}$ in both directions. When it became apparent from cloning the H 2 operon into *E. coli* by way of a lambda phage vector that the switching involved a segment called PD closely linked to the H 2 operon (SILVERMAN *et al.* 1980) the question arose as to whether the *vh 2* genetic locus was the same as *PD*. *PD* would in a sense be analogous to *Ds* in maize and *vh2* analogous to *Ac*. *PD* and *Ds* are, in both cases, cis acting units, *i.e.,* they must be adjacent to the unstable locus to have an effect, but *vh 2* was found to be transacting, *i.e.,* it could act from unlinked sites and must, therefore, code for a diffusible substance which has been identified as a small protein. IINO and KUTSUKAKE (1980) found that the phase-2 stable (*vh2⁻*) *Salmonella* strains could switch to phase 1 (H 1) when prophages P 1, P 7

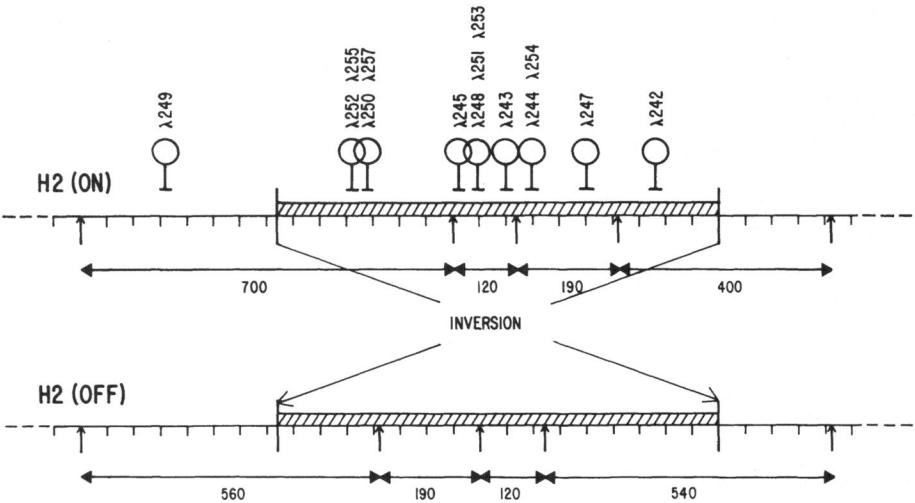

Fig. 7 *A*. To study the phase changes in flagellin, the gene and a flanking sequence was transferred to λ phage cloning vehicle. The H 2 gene (flagellin) with its adjacent vh 2 (control) region was then transferred to *E. coli* which is monophasic. The *E. coli* strain with the integrated λ hybrids then alternated between nonflagellate (off) and flagellate (on) phenotype depending on the orientation of flanking invertable region. The map shows the position of the invertable region (*Hin* gene) and its relation to the H 2 (structural gene) region. (From SILVERMAN *et al.,* 1980; copyright by Cold Spring Harbor Laboratory, 1980)

and Mu were introduced into the cells, *i.e.,* the phage genomes carry a genetic factor similar to *vh2*. The *vh 2* function is also coded by certain bacterial plasmids which could be transferred to *Salmonella* cells. The P 1 and Mu prophages had the remarkable effect of increasing the frequency of phase change to about $2 \times 10^{-2}$ which is hundred fold higher than in the standard diphasic strain mentioned above.

SILVERMAN *et al.* (1980) constructed an *E. coli* strain that carried the H 2 region of *Salmonella* and expressed the H 2 phenotype. A recombinant lambda phage was constructed by inserting a 3.75 kb EcoRI restriction fragment from *Salmonella* that carried the H 2 operon. *E. coli* is normally monophasic with respect to flagella. It produces one flagellin coded by a gene *hag* which is homologous to the H 1 gene of *Salmonella. E. coli* cells lacking flagella (*hag*⁻) were lysogenized with the lambda carrying the *PD* segment in the H 2-on configuration. They produced flagella and could be identified because they were sensitive to another phage, Chi, which enters the cell only by attachement to flagella. The derived *E. coli* strain alternated between a flagellate cell type (H 2-on) and a non-flagellate (H 2-off).

This strain of *E. coli* was used to define the limits of the *PD* locus and the included *vh* 2 gene which Silverman *et al.* (1980) refer to as *hin* (H inversion), *i.e., vh* 2 and *hin* are the same locus. By insertion of a transposable element (Tn 5; see the section below for a description of these elements), they were able to produce switching defective mutants and map the functional regions of the *PD* locus and its linkage to H 2 with a precision of 25 bp (Fig. 7 *A*). This was done using restriction endonucleases and gel electrophoresis for separation of fragments which were then transferred to nitrocellulose sheets and identified with 32 P-labelled probes made from the DNA in the region (Southern blot hybridization). The switching region, *PD,* is about 1,000 nucleotides in extent and located close to H 2. The ends of the *PD* segment are essential to switching and a large region within the locus has to be intact. The ends have to be located close to H 2, but the function of the central region can be replaced by a lambda phage or a plasmid. Therefore, the switching mechanism involved ends that were cis acting while the central part between these two ends can be interrupted by a transposon or a deletion and the switching can occur if a complementing phage or plasmid is available in the cell.

To define more precisely the structure of the switching region a segment including the switching region and a part of the H 2 gene was cloned and sequenced. A portion of the sequence is shown in Fig. 7 *B*. The switching gene *hin* which consists of a promoter region and a coding sequence that could produce a protein with 190 amino acids is bracketed by two 14 bp repeats which are inverted with respect to each other. The *hin* promoter includes a part of one repeat. In the "on" configuration, the sequence is a 14 bp repeat, *hin* promoter, structural gene for *hin,* spacer, H 2 promoter, second 14 bp repeat, and structural H 2 gene. In the "off" configuration, the H 2 promoter is displaced upstream where it fails to operate, while the displaced *hin* promoter has the same relation to its structural gene and continues to be transcribed (Fig. 7 *B*).

The switching function has features in common with other transpositions in *E. coli* and related bacteria, but it is independent of the recombination processes involving the Rec A gene. The inverted repeats bracketing the switching gene appear to be designed for inversion by a recombination event involving these loci (Fig. 7 *B*). If the segments were direct repeats the same recombination event should lead to excision of the bracketed segment as a circular form. If one of the repeats were in a small circle and the other in

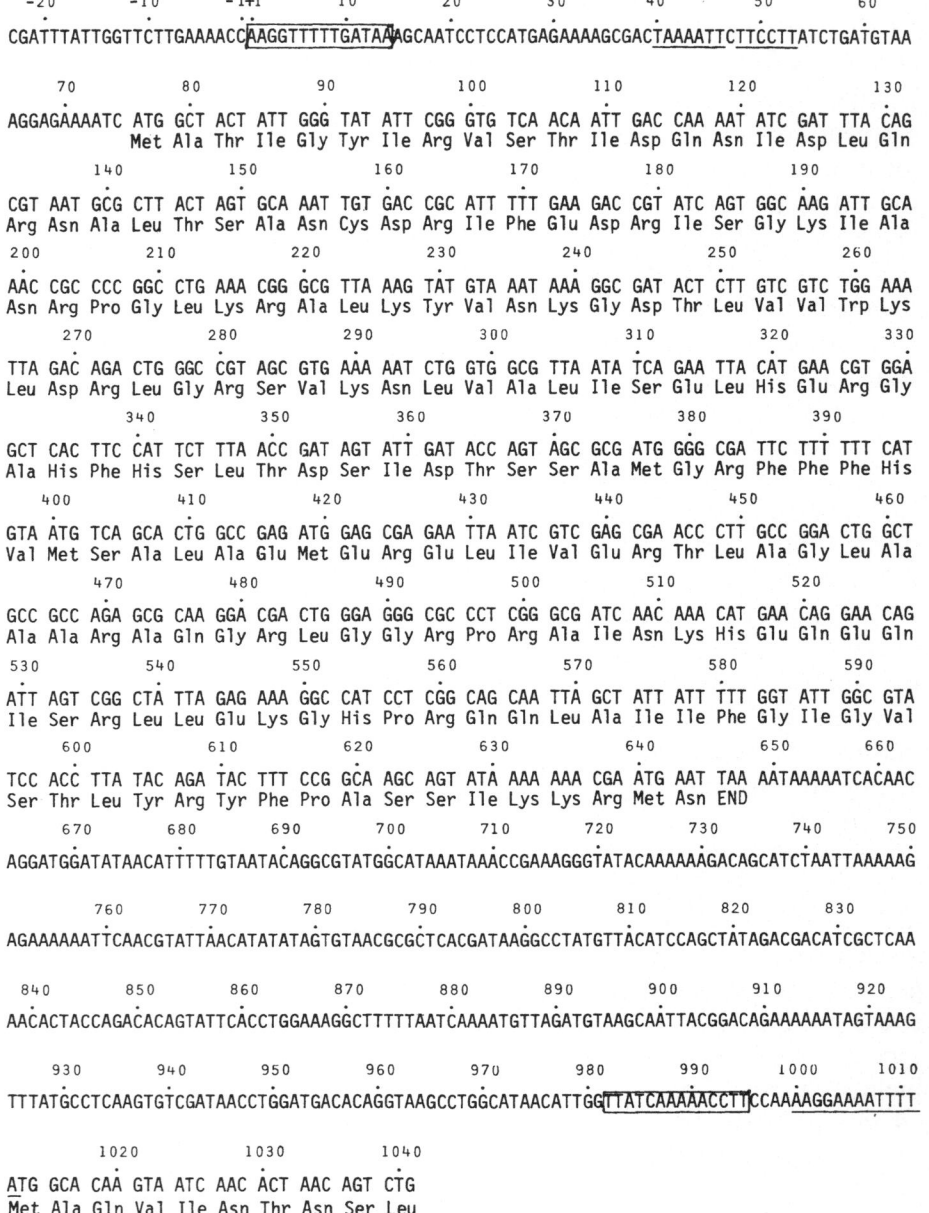

Fig. 7 B. The nucleitode sequence for the switching region (*Hin*). The *Hin* gene is located between nucleotides 76 and 648 and the coding region of the *H 2* gene begins at position 1012 and continues rightward beyond the end of the nucleotide sequence. The inversion region extends from 41 to 995 (5′ to 3′) and is marked by a 14 bp inverted repeat at + 1 to + 14 and + 982 to + 995. The open reading frame for a protein coded by the switching region is shown with predicted amino acid sequence indicated by the DNA base sequence. (From an original photograph provided by M. SIMON, California Institute of Technology, Pasadena)

a large circle or linear DNA (chromosome) the recombinational switch would lead to insertion (integration).

## 2. Insertion Sequences and Transposons

### a) Insertion Sequences

Transposable genetic elements in prokaryotes may be part of plasmids or chromosomes but do not include an origin for replication. This characteristic distinguishes them from replicons. The elements may be classified as insertion sequences (IS) or transposons (Tn). By definition insertion sequences do not carry genes other than those involved in their transposition, *i.e.,* genes coding specialized recombination enzymes thought to be independent of the principal recombination system in *E. coli* that involves the Rec A protein. Transposons (Tn), on the other hand, are defined as larger elements that may code for a transposase, but also carry other genes and may include one or two insertion sequences. The most frequently studied Tn's are those that carry a gene coding for an antibiotic resistance factor, but presumably any gene can become part of a Tn (IIDA *et al.* 1981).

Insertion sequences are bracketed by short direct repeats. The IS 1 from *E. coli,* or plasmids that replicate in these cells, have a 9 bp sequence preceeding and following the 768 bp sequence which comprises the IS (OHTSUBO *et al.* 1980). The mechanism of transposition is not known in detail but may involve the pairing of a portion of the element with a similar sequence at another site. The transposition is thought to occur during replication because during transposition the element is not necessarily lost from its original position (BENNETT *et al.* 1977, SHAPIRO 1979). IS 1 has in addition to the 9 base pair direct repeats, a 35 bp repeat at each end usually in reverse orientation (HEFFRON *et al.* 1979). When transposed it takes one or both of the 35 bp repeats but usually not the 9 bp bracketing segments; a new 9 bp sequence already at the site where it moves will be duplicated, or more descriptively the transposase appears to make staggered cuts separated by 9 bp, the element is ligated to the single chain extensions and the gaps are filled during insertion (Fig. 8 *A*). Insertion elements can be inserted in many places in the genome. They frequently inhibit or restrict transcription especially if inserted in the control sequences of a gene, but in some instances they can enhance transcription. Once inserted they also increase the chance for aberrations, particularly deletions. They may also increase inversions, but these usually are secondary events after the insertion element has moved with a reverse orientation to a second location nearby.

### b) Transposons in Bacteria

An example of a transposon is Tn 3 (Fig. 8 *B*) which codes for three proteins, a transposase, a regulator protein and an enzyme, beta-lactamase which destroys the antibiotic ampicillin (HEFFRON *et al.* 1979). The regulator protein regulates transposase production and its own synthesis as well. This simple transposon has been sequenced and many deletion mutations have been studied in an effort to understand the function of the

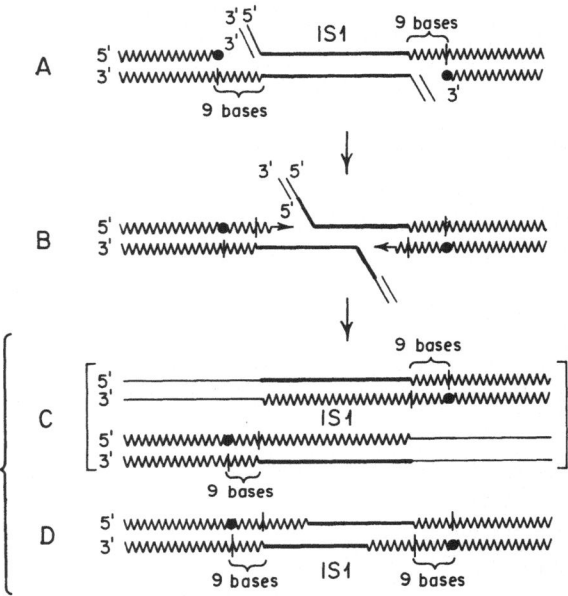

Fig. 8 *A.* A model for the movement of an insertion element (IS) without loss of the element from its original locus. *A.* Intermediate molecule resulting from recombination of a donor plasmid (————) containing the IS 1 sequence (————) with a recipient plasmid (wwww). The donor and recipient plasmids are actually circular, but only portions of them are shown here. Note that a 9-bp sequence in the recipient strands has been subjected to a staggered cut. *B.* Structure formed when displacement DNA synthesis proceeds from the 3′ ends of the staggered cut of the recipient strands in the 5′ → 3′ direction across IS 1. *C.* Cointegrate structure formed when the DNA synthesis proceeds completely across IS 1 and the newly synthesized strands are joined with the 5′ ends of the donor strands. *D.* Recipient molecule, which received an IS 1 insertion. This molecule could be formed from the molecule in B by removing donor strands during displacement DNA synthesis. Note that in both C and D a 9-base duplication has been generated. (Redrawn from OHTSUBO *et al.,* 1980)

three loci (OHTSUBO *et al.* 1980). It has 4,957 bp with the three genes in the order *tnp* A (transposase A), *tnp* R (repressor and resolvase) and *bla* (beta-lactamase). At each end there is a 38 bp inverted repeat (IR-L and IR-R). When all genes are intact the frequency of transposition is low ($4.5 \times 10^{-6}$) per cell per division cycle. Mutants, which inactivate or reduce the effectiveness of the repressor gene, may transpose at a much higher

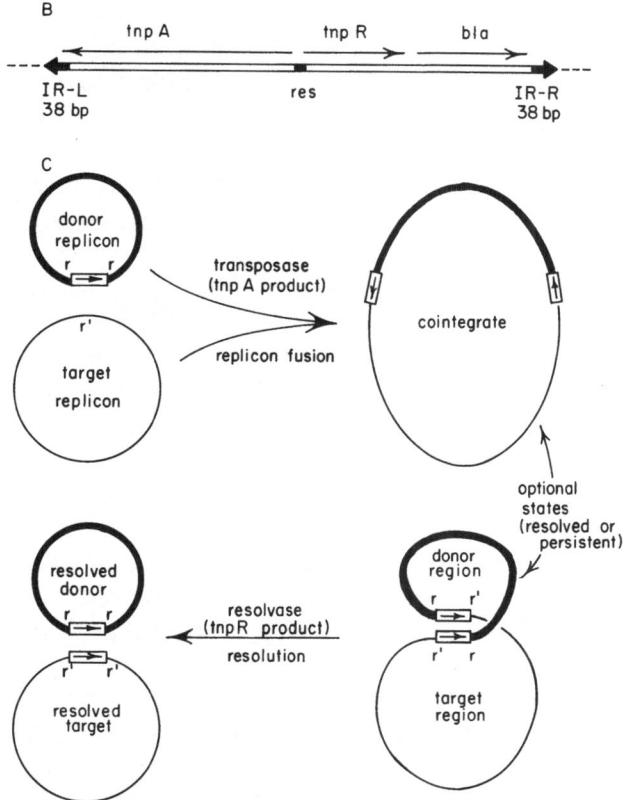

Fig. 8 *B*. Diagram of Transposon 3 showing which shows the three genes, *tnp* A (transposase), *tnp* R (resolvase), *bla* (beta-lactamase) and an internal resolution site (*res*). The structure is bracketed by inverted repeats (IR), left and right and when inserted in a replicon it would have 5 base pair short direct repeats at the ends

Fig. 8 *C*. A generalized model for transposition. The two replicons (plasmid and linear or circular chromosome) are fused by the replication of the transposon, its insertion into the target replicon by a unknown process which has been described as a displacement replication or a roll-in replication. The fusion is catalyzed by the transposase (product of the *tnp* A gene). The fusion product called a cointegrate can exist for a long time or in the presence of an active product of the tnp R gene, a resolvase, it would be separated by recombination. In the case of Transposon 3 the recombination would occur at the resolution site and would be catalyzed by the product of the *res* gene (resolvase). Each replicon would then contain a copy of the Tn 3 bracketed by short direct repeats (r or r′)

frequency, but deletions in one of the IR sequences can reduce transposition.

Transposon 9 is flanked by IS 1 (MacHattie and Jackowski 1977) and new transposons have been generated by using IS 1 (Reif and Arber 1980). Such transposable elements have the 9 bp direct repeats at each end typical of the isolated IS 1. In Tn 9 IS 1 forms the long direct repeats at the ends. However, one copy of IS 1 is sufficient to make a region movable by co-integration. If another element, chromosome or plasmid, also has one IS 1, a co-integrate will form by fusion of the two into one continuous genetic unit, probably by recombination. If the IS 1s are in circular genomes, one in each is enough to cause the circles to fuse. There are many variations of transposons in different bacteria but these few examples will provide a basis for considering transposons in eukaryotes.

The transposition is a two-step process. The first step involves the replication of the transposon coincident with the fusion of the two replicons (Fig. 8 C). The replication may occur by an attachement and displacement replication as shown for an IS unit in Fig. 8 A. After fusion the product called a cointegrate consist of two copies of the transposon in the larger unit. The point of fusion and integration of the donor replicon is determined by the presence of a 5 bp sequence indicated by r (Fig. 8 C). Each transposase may be specific for a certain repeat (r) which may vary in length and sequence for various transposases. The product of gene *tpn* R is required for the resolution; hence the name resolvase. If recombination occurs during resolution in a specific sequence which is a direct repeat, the results will be separation of the replicons into the original donor and target. The same type of separation would occur if the transposon has a single resolution site as shown in Tn 3 (Fig. 8 B). Both would now have the transposable element (Tn 3 in this case) and the 5 bp repeat has been duplicated so that it brackets each transposon (GRIDLEY 1983, and KORNBERG 1982). If the resolvase is defective the fusion product or cointegrate may persist for an indefinite period.

## c) Transposons in Eukaryotes

### 1. A Transposable Element in Yeast, Ty

Ty 1 is a family of similar sequences dispersed in the yeast genome (CAMERON et al. 1979). It is a 5.6 kb segment flanked by 250 bp direct repeats. It is also bracketed by 5 bp short direct repeats which may differ at each locus that Ty 1 is found. Strains of yeast vary in the number of copies of Ty 1, from a few to 30–40. The sequences are transcribed, but the long transcripts of 5,500 nucleotides have no known function. However, Ty 1 can serve as a transposon for genes which are transferred into it. For example, the *his*-3 gene was expressed in some cases when integrated into a Ty 1 element and used to transform yeast cells (SCHERER and DAVIS 1981). The role of Ty 1 in controlling the expression of neighbouring sequences is less clear. Transposition of the element upstream from genes can cause inactivation (CHALEFF and FINK 1980), but it can also cause overproduction of a gene product (ERREDE et al. 1981). It also causes gross genetic rearrangements and increases gene conversion in some positions in the genome.

## 2. Transposable Elements in Drosophila

The moderately repetitive DNA of *Drosophila* occurs as dispersed single copies, 5–7 kb in length separated by longer segments of non-repeated DNA. Three families of such repeats, which have been studied extensively, resemble the Ty 1 element of yeast in overall structure but share little if any homologous base sequences. As examples of these elements, we will consider copia and two similar elements known as 412 and 297 (RUBIN *et al.* 1980). The repeats are present in approximately 30 copies, each at widely scattered sites in the genome. There are many other similar dispersed, repeated sequences which together make up 5–6 percent of the total genomic DNA. The repeats, copia 412 and 297, are bracketed by direct repeats of 0.3 kb, 0.5 kb and 0.4 kb, respectively. They undergo transpositions as shown by the variations in location and copy number when different genetic strains are examined or when cells in culture and embryonic DNAs are compared. At different positions, each is flanked by direct repeats of 4–12 bp which have been duplicated at the site during integration. These three repeats in *Drosophila* and many of the other dispersed repeats are transcribed into long RNAs (YOUNG and SCHWARTZ 1980). Two major classes of these long repeats are about 5 kb and 2 kb in length. Many of the RNAs transcribed from them are capped and polyadenylated and at least one, the 2 kb class, is a functional mRNA as judged by its ability to direct the synthesis of a 50 kd polypeptide in vitro (RUBIN *et al.* 1980). The amount of transcription of copia varies during development with larval stages having the highest level, but if any proteins are translated from these RNAs they have no known function.

Recently mutations at the white locus have been correlated with insertion of small duplications (COLLINS and RUBIN 1982, and KARESS and RUBIN 1982). A mutation of the wild type allele $w^+$ to $w^i$ (white-ivory) is unstable with a mutation rate of $5 \times 10^{-5}$ in homozygous females and $5 \times 10^{-6}$ in males and deletion heterozygous females. Cloning of the wild type white locus and the mutants has provided the tools for analysis of the locus by restriction enzyme digestion and Southern blot analyses. The analyses showed a tandem duplication of 2.9 kb of the DNA at the locus. Reversions appeared to have the duplicated segment removed by excision at a point that removes the exact length. Two derivatives of the locus, one wild-type (phenotype) and the other with a partial phenotypic change, carry insertions of moderately repetitive DNA from outside the locus along with deletion of some sequences from the wild type locus.

In a derivative of the white-ivory mutant (COLLINS and RUBIN 1982) the change involved the transposition of a moderately repetitive sequence into the region. One fragment of 2.9 kb is missing and two new fragments of 5.5 kb and 7.5 kp appear. This is a net gain of 10 kb at the locus and gives an indication that a mobile element has been inserted.

## 3. Retroviruses

Vertebrates cells harbor DNA copies of RNA viral genomes which if integrated into the host genome can be dormant except for replication along with the chromosomal DNA. Some of these proviruses can become active viruses in some cells during development. RNA of the infecting viruses is reverse transcribed and the copy which is double stranded DNA is then integrated into the host cell genome. The DNA copies of this group of viruses have structural properties in common with the Ty 1 elements of yeast and the copia-like elements of *Drosophila*. All are similar to mobile genetic elements with direct or inverted terminal repeats which could move as part of the element. The retroviruses cause duplication of 4,5 or 6 nucleotide pairs at the target site of integration. However, some distinctions between retrovirus and copia and Ty 1-like elements are evident; for example, the efficiency of retrovirus integration is greater and the retroviral genome is trimmed of a few nucleotides at its end before integration (SHOEMAKER *et al.* 1980). Retroviruses are not known to undergo transposition, but appear to be "fixed" at the site of integration. In this respect they may have evolved some specialization from an ancestral copia-like element to make them more efficient for integration. Their reproductive cycle requires integration as a regular event.

All of these elements, Ty 1, copia-like elements in *Drosophila* and retroviruses of vertebrates share a common sequence at the site of integration. The first two nucleotides at the 5' end of the element is the dinucleotide TG and at the 3' end CA. This suggests that the nicking enzymes which functions in their integration may have a common evolutionary origin.

## 4. Alu Family of Sequences in the Human Genome

The moderately repeated sequences of mammals include short sequences that are dispersed throughout the genome. These vary from 130 bp to over 300 bp in length and have some of the characteristics of transposable DNA segments (reviewed by SCHMID and JELINEK 1982, and JELINEK and SCHMID 1982). In the human genome a large class of these dispersed repeats have a single Alu I restriction endonuclease site about 170 nucleotides from the 5' end (Fig. 9). The end is defined as the beginning of the transcript made by RNA polymerase III which will transcribe many of the sequences in vitro. Th dispersed repeats of rodents are shorter than the Alu family of the human genome; they are only about 130 bp. The human Alu family can be considered as a direct repeat of an ancestral type similar to those in the

```
     direct repeat     ↓          ↓          ↓          ↓          ↓
-------------------↓                                                        50
...AAGATTCACTTGTTTAGAGGCTGGGAGTGGTGGCTCACGCCTGTAATCCCAGAATTTTGGGAGGCCA



        ↓          ↓          ↓          ↓          ↓          ↓
                                                                       120
AGGCAGGCAGATCACCTGAGGTCAAGAGTTCAAGACCAACCTGGCCAACATGGTGAAATCCCATCTCTAC
                        └─────────────┘
                        RNA polymerase III
                          promoter region

        ↓          ↓          ↓          ↓      ←0→              ↓
                                                 →                    190
AAAAATACAAAAATTAGACAGGCATGATGGCAAGTGCCTGTAATCCCAGCTACTTGGGAGGCTGAGGAAG
                                                 └──┘
                                              Alu sequence


        ↓          ↓          ↓          ↓          ↓          ↓*******↓260
GAGAATTGCTTAAACCTGGAAGGCAGGAGTTGCAGTGAGCCGAGATCATACCACTGCACTCCAGCCTGGG



        ↓          ↓          ↓          ↓          ↓    direct repeat
                                                        ↓------------------
TGACAGAACAAGACTCTGTCTCAAAAAAAAAAAAAAGAGAGATTGATTGAAAAGATTCACTTGTTTAG...
```

Fig. 9. The Alu sequence which is positioned 5′ to the Gγ gene in the human β-globin cluster. Features typical of Alu sequences are indicated. The Alu sequences are typically bracketed by short direct repeats of 8 to 20 nucleotides (this copy is bracketed by a 17 bp repeat overlined with a broken line). Some of the sequences are transcribable by RNA polymerase III; the suggested internal promoter site is indicated. The name for these dispersed repeats comes from a conserved Alu sequence indicated as the zero position, since BLUR clones were sequenced in both directions from the end produced by the Alu restriction endonuclease. The sequence is considered to be derived as a dimer of a primitive Alu-type sequence with a 31 bp insert shown by underlining. A highly conserved region is shown by the asterisks. This sequence is conserved in all mammalian sequences so far determined and is present in DNA replication origins of several papavoviruses, including SV 40. The double underlined sequence, GAGGC, is also found repeated in the origin of SV 40 where it appears to be involved in binding T antigen (DE LUCIA et al., 1983). The Alu sequences typically end in clusters of adenines which may be seen from sequence 121–133 and 283–295, since the human Alu sequence is a dimer.
(Adapted from DUNCAN et al., 1980)


rodent genome. It has evolved with base pair mutations that now make the two direct repeats less homologous than some similar sequences in related species. In addition to the duplication there has been a 31 bp insert about 83 bp downstream from the beginning of the second copy. Just following this 31 bp insert is a 9 bp highly conserved sequence that is the same in all mammalian Alu-like sequences so far examined for this property.

The Alu sequences in the human genome and the Alu-equivalent sequences in other vertebrates have these properties in common: 1. they are flanked by direct repeats 8–20 bp in length that were presumably derived from the genomic target sites just as in the case of insertion sequences, transposons, Ty 1 segments, copia-like segments and retroviruses; 2. the 3′ end has an adenine-rich sequence that is not conserved among the different Alu or Alu-equivalent sequences and the 3′ ends are less precisely defined than the 5′ end where the RNA transcripts are initiated; and 3. the promotor region is internal and for the Alu family has been tentatively identified as the sequence GAGTTCPuAGACC in the first monomer unit. The variability in the length of the flanking sequences indicate that the enzymes functioning in transposition may be related proteins, but specific for certain sequences. However, no correlations have yet been made between heterogeneity of the sequence and the length of the direct repeats and, of course, there may be none. The sequences are constant enough around some known genes in related species to suggest that they move infrequently. However, there are also enough variations to suggest that transposition does occur (see the pattern for the hemoglobin gene cluster below). There are about 300,000 copies of the Alu family in the human genome which means that there would be one every 6–8 kb if they were evenly distributed. The ratio of those that are transcribable by RNA polymerase III to those not transcribed is uncertain, but of the eight that have been found in the cluster of beta-like hemoglobin genes in a region of about 50 kb, only 6 are transcribed and these are spaced at intervals of 8–12 kb (FRITSCH *et al.* 1980).

The functional role of these sequences is unknown, but the fact that they have sequences in common with the DNA replication origins of papova viruses, such as SV-40, polyoma, and human BKV leads to the suggestion that they may be origins of replication. There is some evidence to support this idea (TAYLOR and WATANABE 1981, and CHAMBERS *et al.* 1982), but more data will have to be obtained to confirm the hypothesis. Even if they serve as origins for replication and transcription they may also have alternate or supplementary roles as regulatory sequences. As mentioned above they are not only transcribed in vitro, but at least some are transcribed in vivo. There are species of nuclear 4.5 S RNA that are homologous to the Alu sequences and the cytoplasmic 7 S RNA is partially homologous with the sequences. Some of the Alu sequences are found in the heterogeneous nuclear RNAs where they are transcribed in the introns and usually spliced out in the formation of mRNAs. In vitro transcripts are usually short and end at one or more distinct sites about 400 to 500 nucleotides from the origin, *i.e.* downstream from the 3′ end of the Alu-equivalent sequences.

Although the Alu sequences are the major family of dispersed sequences in human cells, other moderately repeated sequences have been detected and it has been estimated that one could occur about every 2,000–3,000 nucleotides in most of the DNA. Other mammals have a comparable number but those in most other mammals are shorter and perhaps more heterogeneous than the Alu family. However, the investigations are insufficient to rule out sequences the size of human Alu family in some vertebrates.

These sequences resemble transposable elements since they are bracketed by short direct repeats, but they do not appear to code for a transposase or any other protein in spite of the fact that many are transcribed by RNA polymerase III. Their role in methylation, if any, is not demonstrated but they have a high CpG content and many of the mutations have occurred in these doublets which has led to the sequence heterogeneity of Alu family. As will be proposed later, they might serve as recognition sites for methylases as well as proteins involved in replication.

# III. Differentiating Systems and Their Methylation Patterns

## A. Differentiation of the Hemopoietic System and Organization of the Hemoglobin Genes

The hemoglobin gene clusters provided the first examples of correlations between DNA methylation and gene function at specific loci. The hemopoietic system consists of populations of several types of cells that are derived by the division of hemopoietic stem cells. The time of differentiation of the stem cells during development is difficult to determine because they have few known distinguishing phenotypes except their potential for self reproduction and the generation of cells that can differentiate into most, if not all, of the cells which reside in the myloid tissues, primarily bone marrow, and the lymphoid tissues in lymph nodes, spleen and the thymus gland (TILL 1981, SIMINOWITCH *et al.* 1963, CURRY and TRENTIN 1967). The cells of the myloid tissues include three major types of precursor cells: 1. proerythroblasts which divide and differentiate into erythroblasts and ultimately the erythrocytes, 2. myeloblasts which divide and form myelocytes (these continue to divide and differentiate into granulocytes), and 3. megakaryoblasts which divide and differentiate into megakaryocytes (when mature the megakaryocytes fragment into the small blood platelets which lack a nucleus and survive a relatively short time).

The erythroblasts and their descendents, the erythrocytes, are the ones which have been studied with respect to methylation of the hemoglobin genes. The erythrocytes are called reticulocytes when young and maturating into functional cells with no further potential for division. In mammals the cells lose their nuclei before maturity and have a life time of a few weeks or months at most. They must be continually regenerated from the stem cells throughout the life of the animal. One stem cell or a few cells can produce a colony of several million cells which are not homogeneous; they produce both erythrocytic and granulocytic cells which are recognizable by their morphology and staining properties (TILL 1981). The stem cells have no distinguishing morphology that allows their recognition under the microscope, but they can be recognized by their ability to produce colonies. If a mouse is given a dose of X-rays sufficient to kill nearly all of its hemopoietic cells (a lethal dose), it can be rescued by injection of a suspension of bone marrow cells into the tail vein. Some of the cells will form colonies in the spleen and the number of colonies is related to the number of cells injected. Those cells which divide and produce large colonies after 14 days contain cells of more than one hemopoietic linage as well as stem cells capable of regenerating multilineal colonies if injected into irradiated mice (MAGCI et al. 1982).

Red cells are usually programmed to produce a single type of hemoglobin at any one developmental stage. Higher vertebrates have two or three classes of hemoglobin molecules in red cell populations at different times in development. The first red cells produced in the early human or mouse embryo have a type of hemoglobin called E. As the embryo grows, cells which produce hemoglobin F (fetal) become abundant in the population and as the hemoglobin E producing cells die, they are not replaced. In the fetus a large population of reticulocytes begins to produce the adult hemoglobin A, which at birth is about 20 percent or more of the total. Normally, it completely replaces the fetal hemoglobin within the first year of the human infant's life. Human hemoglobin A molecules are constructed of four polypeptide chains, two identical alpha and two identical beta chains, which bind two heme groups. All other hemoglobins are similar and have two identical alpha-like chains and two beta-like chains.

During development there are some exceptions to the generalization that each cell produces one class of hemoglobins, for example WONG et al. (1982) reported that circulating peripheral blood cells and disaggregated cells of the yolk sac of a mouse embryo at day 9 of gestation already have a population of cells which are programmed to synthesize adult hemoglobins. At day 9 the three embryonic hemoglobins (E I, E II and E III) are predominant and are synthesized in nucleated erythrocytes produced in the blood islands of the yolk sac. In retaining their nuclei, these cells resemble the erythrocytes of the ancestral vertebrates, birds, reptiles, amphibians and fishes. These primitive red cells of the mouse embryo later synthesize and accumulate adult hemoglobin in addition to the embryonic types. By day 12 the embryonic liver becomes the active organ of erythropoiesis and continues to be the major site until near birth. The erythrocytes produced in fetal liver are non-nucleated and contain only adult hemoglobins. These different cells should provide interesting objects for studies of changes in methylation patterns, if techniques will allow their isolation.

## The Beta and Alpha Gene Clusters

The genes coding for the beta-like chains usually form a single cluster and probably arose in evolution by duplications of an original gene which then mutated independently of the ancestoral gene. The beta gene cluster of man is now arranged over a region of about 50 kb as shown in Fig. 10 A (FRITSCH et al. 1980).

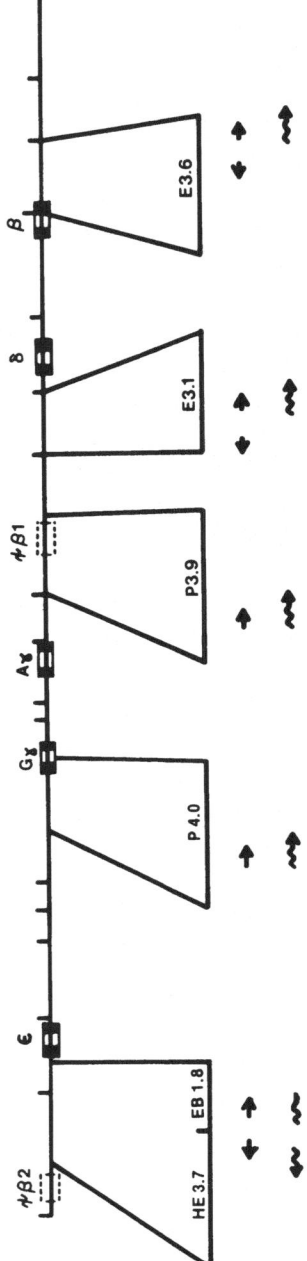

Fig. 10 A. The seven β-like genes of the human β cluster are arranged in a sequence on the chromosome over a region of about 50 kb. The functional genes are all transcribed to the right from the ε (epsilon) gene which codes for the very early E hemoglobin in the embryo to the β (beta) gene which codes the major adult hemoglobin. RNA polymerase III transcripts hybridize to the regions indicated by the horizontal wavy arrows. These regions, as well as those indicated by the straight arrows, represent Alu sequences. Note that the sequence given in Fig. 9 is the Alu sequence that flanks the Gγ gene. (From FRITSCH et al., 1980; copyright by Cold Spring Harbor Laboratory, 1980; photograph provided by Dr. CHE-KUN JAMES SHEN)

There are five functional genes and two pseudogenes which are still carried in the cluster but no longer code for a protein. The gene on the extreme left (the orientation is the direction of transcription which is from left to right) is a pseudogene, $\psi\beta 2$, and the next is the epsilon gene which codes for the beta-like chain of hemoglobin E in the very early embryo. Then comes two gamma genes that function coordinately to code for the gamma chains of fetal hemoglobin. Next in order is another pseudogene, $\psi\beta 1$, and then another coordinate pair, delta and beta, that functions in the adult to code for the beta-like chains of the two adult hemoglobins. However, very little of the total hemoglobin is coded by the delta gene.

The beta-like clusters are similar in the various mammals so far investigated. All code for at least two types of globin chains, one or more embryonic types and one or more adult types.

The alpha-like chains of hemoglobin are nearly as variable as those in the beta cluster and most of them are coded by genes, which, like the beta cluster, are arranged in the sequence of their expression during development.

### The Alu Family Repeats in the Human Beta Cluster

Because of some suggestions that the Alu repeats and similar dispersed repeats may have some regulatory role in gene function (BARALLE et al. 1980, and BELL et al. 1980), the distribution of Alu repeats in the human beta cluster will be considered in so far as it has been analyzed. Fig. 10 A shows the beta cluster and the reported Alu sequences (FRITSCH et al. 1980). The location was determined by subcloning various segments of the 50 kb region and hybridizing with a labeled repeat after the fragments produced by restriction endonucleases were separated on a gel by electrophoresis and transferred to nitrocellulose filters. The orientations of the repeats are shown by arrows and were determined as follows: 1. by the direction of the RNA polymerase III transcripts, if a transcript is made in vitro; 2. by the orientations indicated by electron microscope analysis of heteroduplexes and loop and stem structures in single chains; and 3. by the orientation of the repeats from sequence analysis (SCHMID and JELINEK 1982). The sequences are transcribed from the first nucleotide 3' to the first direct repeat which usually brackets the Alu repeats. The 3' end of the sequences are marked by clusters of A's that are not highly conserved. The in vitro transcripts run 150–200 nucleotides beyond the end indicated by the cluster of A's and usually terminate in a cluster of T's preceded by a G:C rich region. Two features may be worth consideration: 1. the orientation of the transcribable sequences are all in the same direction as the RNA polymerase II transcripts of the structural genes, and 2. it is possible to relate one transcribed Alu sequence to each functional unit or coordinately controlled pair of genes. If the upstream transcribed sequence is the significant one, there is one for the epsilon gene, one for the coordinately controlled gamma pair, one for the beta 1 pseudogene and one for the coordinately controlled delta and beta genes. The small amount of data available has been interpreted to indicate that the Alu sites 3' and perhaps 5' to the gene are somehow involved in regulation (FRITSCH *et al.* 1980). Several deletions in the delta-beta region of the cluster are correlated with the failure of the gamma genes to be turned off at the usual time in development. It should be emphasized that there is, as yet, no convincing evidence for a functional relationship such as the one predicted above on the basis of the occurrence of a transcribable Alu site to the 5' side of each functional gene or coordinately controlled gene pair. However, the suggestion which I make is interesting because of the evidence that transcribable Alu repeats may be origins of replication (CHAMBERS *et al.* 1982) and changes in methylation patterns appear to be correlated with replication. If these four Alu sequences are origins for unidirectional replication, the leading strands should be identical to the transcript and the retrograde chain formed from OKAZAKI fragments should be on the non-transcribed chain (TAYLOR 1983). If the maintenance methylase attaches at the primer site of the origin and moves to the right until it reaches another origin, it would be expected to maintain the methylation pattern for one unit of replication per binding site. An inhibitor molecule attached to an origin would protect every site in the unit and lead to a deletion of the methylation in such functional units without affecting the preceding or the following replication units. This hypothesis assumes that the origins (presumably Alu sequences) have enough sequence specificity and variation from site to site to influence the binding of the methylase, perhaps before the primer is removed following replication (see section on maintenance and deletion of methylation patterns). Such a mechanism would prevent the methylation of only one of the two chromatids because of the orientation of replication and the primer. One chromatid, the one replicated by OKAZAKI pieces in the retrograde mode, would retain the original pattern and therefore could maintain a stem cell.

## B. Methylation Patterns of Hemoglobin Genes

### 1. A Method for Locating Single CpG Sites by the Use of Two Restriction Endonucleases Which Are Isoschizomers

The method that has been most useful for analyzing the methylation patterns of genomic DNA was described by BIRD and SOUTHERN (1978). Since it was known that most of the methylcytosine in DNA of vertebrates is found in CpG doublets, the presence of 5 mC could be detected by the use of two restriction endonuclease (isochizomers), that share a recognition sequence containing CpG. One must be inhibited when the C is methylated while the other is not. The most useful pair yet discovered was the one suggested by BIRD and SOUTHERN (1978) and used by WAALWIJK and FLAVELL (1978). The nucleases are synthesized by different genera of bacteria. Hpa II is found in *Hemophilus parainfluenza* and Msp I in an unnamed species of *Morexella* sp. Both Hpa II and Msp I cleave DNA at CCGG sites, but Hpa II is completely inhibited if the inner C is methylated at the 5 position in the pyrimidine ring. Msp I is not inhibited by methylation at the inner C, but was later shown to be inhibited if the outer C is methylated (KAPUT and SNEIDER 1979, and SNEIDER 1980).

With these two endonucleases one can first cleave a segment from the genomic DNA with another restriction endonuclease which makes long segments, for example, Eco RI. These Eco RI segments are separated by electrophoresis in an agarose gel. The particular segment of the gene and its flanking regions are identified by transfer of the segments to a nitrocellulose filter by the Southern blotting procedure and hybridizing with labeled cloned segments of the gene. If the genomic DNA is further digested with Hpa II and Msp I the methylation pattern in CCGG sites will be revealed. Suppose the gene is not cleaved by Eco RI but the DNA is cut at either end of the gene. In the lane run without further digestion, a single band hybridizes with the labeled probe (cloned segment). If the Eco RI segment containing the gene has three CCGG sites, the DNA digested with Msp I will usually separate into four bands that bind the labeled probe. Let us suppose that one of the sites is regularly methylated in genomic DNA; then in the lane containing the Hpa II digest only three bands will appear. If a restriction map of the gene is available, the number and location of the methylated cytosines in the CCGG sites can be definitively located by analyzing the Southern blot of these three samples.

Although isoschizomers suitable for other sequences are unknown in most instances, there are several other restriction enzymes whose recognition sequences contain CpG sites and are useful for identification of methylated CpG sites in genomic DNA by methods similar to that described above (MCCLELLAND, 1983). The most useful of these restriction enzymes aside from the Msp I-Hpa II pair is Hha I, which recognizes and cleaves the GCGC site unless the inner cytosine is methylated. The isoschizomers Sma I and Xma can be used for identifying methylation at the inner cytosine in the CCCGGG sequence in a similar way to that illustrated above for the Msp I-Hpa II pair, but this sequence is rare.

## 2. The Genes of the Beta Cluster Which Function in a Reticulocyte Are Undermethylated

WAALWIJK and FLAVELL (1978) used the restriction enzyme pair, Hpa II and Msp I, to investigate the methylation of a CCGG site in the intron of the beta-globin gene of the rabbit. They were unable to isolate a pure population of erythroblasts but they looked at the DNA of erythroid tissues and other somatic tissues where they found 50% of this CCGG site methylated. The same CCGG site in the sperm DNA was 100% methylated and the site in brain cell DNA was 80% methylated. Although this did not provide a definitive correlation of undermethylation with function, the expected difference was in the right direction. There were at least two defects in the system analyzed. The cell populations analyzed were heterogeneous, because a pure population of erythroblasts and reticulocytes was not available. Since the CCGG site was in an intron, it would, if unmethylated, probably not be the critical site. However, its absence might serve as an indicator of the deletion of a methylation pattern in other parts of the gene.

A more extensive analysis was made of the beta cluster in the human genome by VAN DER PLOEG and FLAVELL (1980). They used cloned gamma and beta cDNAs from plasmids as the probes for the hemoglobin genes, gamma, delta and beta in Southern blots. The probes were labeled by nick translation with $^{32}$P-nucleoside triphosphates. The DNA from the different types of cells was cut with appropriate restriction endonuclease to release large fragments containing parts of the genomic cluster. These fragments were further digested with Hpa II and Msp I to locate the methylated CCGG sites.

The sperm DNA from 3 individuals was methylated at all sites detectable (17 sites cut by Msp I). They also used other endonucleases inhibited by methylation of a CpG doublet in the recognition site without finding any change in band patterns in the Southern blots. Since these enzymes have no known isoschizomers which cut the sites when methylated, this part of the analysis gives little information except that most of the CpGs are probably methylated in the DNA of the beta cluster of sperm.

In somatic tissues not expressing the globin genes the CCGG sites in the globin cluster was usually but not always highly methylated. Fetal brain and lymphocyte DNA was methylated at most of the seventeen sites, but placenta and two cell culture lines, HeLa and KB, were very much undermethylated. Adult blood cells (white cells) and lymphocyte cell lines on the other hand had high levels of methylation in the beta cluster. Cells from fetal liver where the gamma genes are expressed (50% erythroid cells) have the CpG unmethylated in a major component and a high level of modification in a minor component of about 25% of the DNA.

VAN DER PLOEG and FLAVELL also examined a cell line, K 562, which can be induced to express the epsilon and gamma globin genes, but not delta or beta. The gamma region was almost completely unmethylated in both the induced and uninduced cells. The delta and beta regions were highly methylated and did not change upon induction of the epsilon and gamma genes.

One can summarize the findings by paraphrasing the statements of the authors. Undermethylation appears to be a necessary condition for expression of the genes, but undermethylation alone is not sufficient for expression as shown above for HeLa and KB cell lines. Many examples are now available to show that the genes are not necessarily expressed when the methylation is deleted, but undermethylation appears to be a necessary condition before other regulatory steps leading to expression can occur.

### 3. Tissue Specific DNA Methylation Occurs in the Cluster of Rabbit Beta-Like Globin Genes

DNA methylation in the beta-like cluster of globin genes of rabbits was studied in the CpG sites of the CCGG sequences by the use of Hpa II and Msp I endonucleases (SHEN and MANIATIS 1980). There are four genes in the cluster and four cloned fragments containing mainly the large introns of the four genes were available to label with $^{32}$P and use as specific probes for the different genes. The probes were labeled by nick translation and hybridized to the cleaved genomic DNA which had been transferred by blotting to nitrocellulose filters after separation according to size by electrophoresis in 1% agarose gels. The analysis of beta 3 is illustrated in Fig. 10 B (SHEN and MANIATIS 1980). There are 13 HpaII sites in the region of the four genes studied and those around beta 3 are shown to be sites 4, 5, 6 and 7. Site 6 is at the 5' end of the gene; 4 and 5 are far upstream and 7 is beyond the 3' end. Site 6 was totally methylated in all non-expressing tissues examined including bone marrow. Only in the cells of the blood island where it is expressed was it partially unmethylated.

In Fig. 10 C the results from all of the probes are summarized. Three genes have a single Hpa II site that is fully methylated in non-expressing tissues, but there is no site close to beta 2 to indicate its state of methylation. Genes beta 3 and beta 4 have good indicator sites and they are fully methylated in non-expressing tissues.

Since only a small fraction of the CpG sites is detectable by this method, one may not be seeing the most important sites for regulation. The best indicator is probably site 6 in the beta 3 gene, since it is located near the 5' end of the coding region.

The beta cluster of chicken globin genes consist of four functional genes (LITTLE 1982), two in the embryo and two in the adult. Unlike the human and mouse genes, these are not in the sequence used in differentiation. The two genes for adult hemoglobin are bracketed by the two that function in the embryo. No pseudogenes have been discovered in or near the cluster. McGHEE and GINDER (1979) made a preliminary examination of the methylation pattern by the use of the Msp I-Hpa II endonuclease pair and Southern blots. The advantage over the studies with mammals was that relatively pure preparations of nucleate erythrocytes could be examined, but because the probe used, a cDNA clone of the adult chicken beta-globin gene, hybridizes to some extent with the embryonic beta-globin sequence, the results were not as clear as would be desirable. The tissues in which the beta-globin gene is not expressed (oviduct, brain and embryonic red blood cells), have the Hpa II sites at least partially methylated. In erythrocytes and reticulocytes from adults where the adult beta-globin gene was being expressed or had been, the Hpa II sites were unmethylated.

### 4. The Alpha-Globin Gene Cluster in the Chicken and Xenopus

The alpha-globin gene cluster in the chicken has been examined for methylation by HAIGH *et al.* (1982). There are 3 genes in the cluster arranged in the order pi, alpha D and alpha A. The pi gene is expressed only in embryos and the alpha genes code for the adult hemoglobins. The Hpa II-Msp I and the Hha I sites were mapped in the region of the genes and for some distance in the region 5' to the genes. The sperm DNA is totally

---

Fig. 10 *B*. Methylation of CCGG sites flanking the embryonic β 3 globin gene of the rabbit. DNA from yolk sac blood islands and bone marrow was digested with Hpa I, Hpa II, Msp I or combinations and hybridized to $^{32}$P-labelled β 3 intron probe (the white space in the diagram of the gene shows the position of the intron). *a* Autoradiograph showing the hybridization pattern which demonstrates the degree of methylation of the Hpa II sites, 4, 5, 6, and 7 in blood islands (lanes 1–5) and bone marrow DNA (lanes 6–10). Lanes 1 and 6 were digested with Hpa I which cuts out a segment of 6.9 kb which includes Hpa II sites 4, 5, and 6; lanes 2 and 7 digested with only Hpa II; lanes 3 and 8 with only Msp I; lanes 4 and 9 with Hpa I + Hpa II; and lanes 5 and 10 with Hpa I + Msp I. Scales on the sides of the autoradiograph are sizes of fragments in kb. *b* Map of CCGG sites; the fragments produced by methylation patterns at various Hpa II (Msp I) sites and the expressions for estimating their methylation frequencies (*f*). The fragment lengths are indicated above each line representing a fragment and these correspond to the fragments observed in the autoradiograph as a dark spot on the film (sizes on the right side of the autoradiograph). The methylation frequencies, f 4, f 5, f 6, and f 7 at the corresponding Hpa II sites were calculated from densitometer scans and the measured area under the individual peaks. From the areas the weight proportion, P, of each band within a lane was calculated. Since each lane contained the same amount of DNA the weight proportion of a band is related to the methylation frequency, f, of the Hpa II sites within the fragment. For example, if the 4.9 kb band is to be detected in a Hpa I + Hpa II digest site 6 must be at least partially methylated and site 5 must be partially unmethylated. The probability of detecting the 4.9 kb band or its relative density (*P*) is the product of (1−f 5) and f 6 in which f 5 and f 6 are the methylation frequencies of sites 5 and 6, respectively. (From SHEN and MANIATIS, 1980; photograph provided by Dr. TOM MANIATUS)

(a)

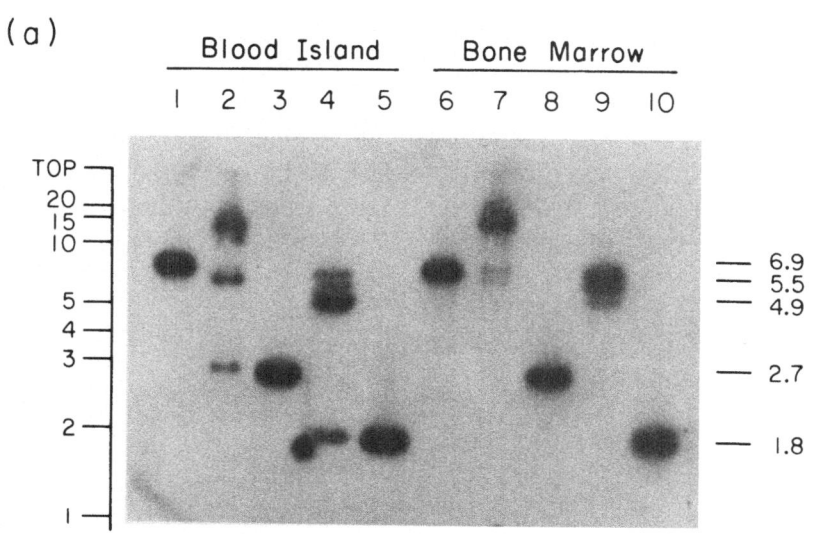

(b)

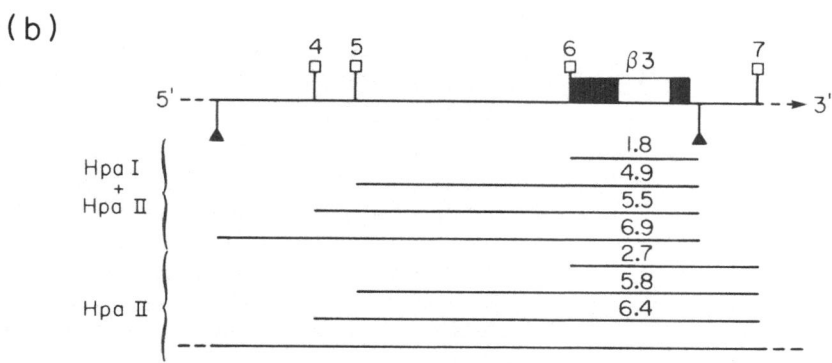

$$( HpaI + HpaII )\quad 1 \cdot f_4 \cdot f_5 \cdot f_6 \cdot 1 = P_{6.9}$$

$$(1-f_4) \cdot f_5 \cdot f_6 \cdot 1 = P_{5.5}$$

$$(1-f_5) \cdot f_6 \cdot 1 = P_{4.9}$$

$$(1-f_6) \cdot 1 = P_{1.8}$$

$$HpaII \qquad (1-f_6) \cdot (1-f_7) = P_{2.7}$$

methylated in the sites in and near the cluster except for one Hha I site at the extreme left of the segment examined and far from the genes. The brain DNA was similarly methylated except that some of the sites near or within the genes were only partially methylated. The DNA in the red blood cells was unmethylated at the near 5′ flanking sites when the gene was functional. The differences in the pi gene and the adult globin genes was most striking. In embryonic red blood cells three sites within and in the 5′ flanking region of the pi gene were unmethylated while these were fully methylated in

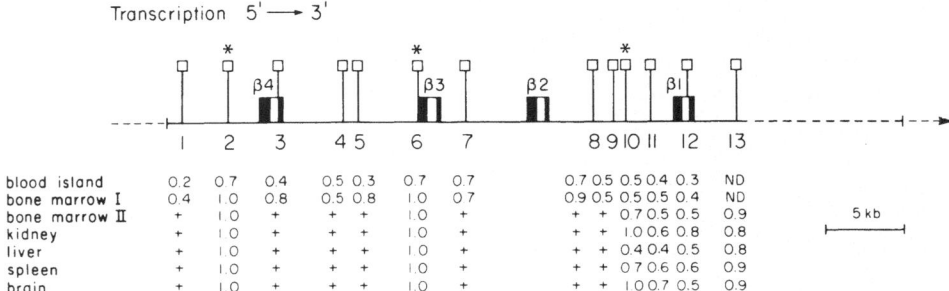

| | 1 | 2 | 3 | 4 | 5 | 6 | 7 | 8 | 9 | 10 | 11 | 12 | 13 |
|---|---|---|---|---|---|---|---|---|---|---|---|---|---|
| blood island | 0.2 | 0.7 | 0.4 | 0.5 | 0.3 | 0.7 | 0.7 | 0.7 | 0.5 | 0.5 | 0.4 | 0.3 | ND |
| bone marrow I | 0.4 | 1.0 | 0.8 | 0.5 | 0.8 | 1.0 | 0.7 | 0.9 | 0.5 | 0.5 | 0.5 | 0.4 | ND |
| bone marrow II | + | 1.0 | + | + | + | 1.0 | + | + | + | 0.7 | 0.5 | 0.5 | 0.9 |
| kidney | + | 1.0 | + | + | + | 1.0 | + | + | + | 1.0 | 0.6 | 0.8 | 0.8 |
| liver | + | 1.0 | + | + | + | 1.0 | + | + | + | 0.4 | 0.4 | 0.5 | 0.8 |
| spleen | + | 1.0 | + | + | + | 1.0 | + | + | + | 0.7 | 0.6 | 0.6 | 0.9 |
| brain | + | 1.0 | + | + | + | 1.0 | + | + | + | 1.0 | 0.7 | 0.5 | 0.9 |

5 kb

Fig. 10 C. A map of the β 4, β 3, β 2, and β 1 genes of the β-like gene cluster of the rabbit along with the methylation frequences at 13 Hpa II sites calculated as illustrated in Fig. 10 B. The positions of the Hpa II sites (CCGG) are shown by the square white boxes above the map. The * on top of sites 2, 6, 10 indicates that these sites are totally methylated in non-expressing tissues. The genes are represented by black (exon sequences) and white (intron sequences) boxes. Below the map is a table of the methylation frequencies at each of the 13 Hpa II sites in different tissues (bone marrow I and II are from different rabbits; the blood island DNA and bone marrow I were from the same animal). The + symbol means that the methylation frequency at a specific Hpa II site is approximately the same as that in bone marrow I DNA, but the calculation shown in Fig. 10 B was not carried out. (From SHEN and MANIATIS, 1980; photograph provided by Dr. TOM MANIATIS)

normal adult red blood cells as well as those of animals made anemic by phenylhydrazine treatment. All but one site was unmethylated in the region of the adult alpha D and alpha A genes in red blood cells from mature chickens. There were a number of apparent exceptions noted where one would have expected to find no methylation if function were correlated with a complete absence of methylated sites within and flanking the gene. Whether these expectations are correlated with the partial methylation within each cell or due to population heterogeneity could not be determined.

One interesting finding was a cluster of Hpa II sites over a distance of 3.5 kb 5′ to the cluster of genes which were resistant to Msp I in the DNA extracted from all tissues, but not in cloned segments. Presumably this represents methylation of the first cytosine in the CCGG sites. Similar sites

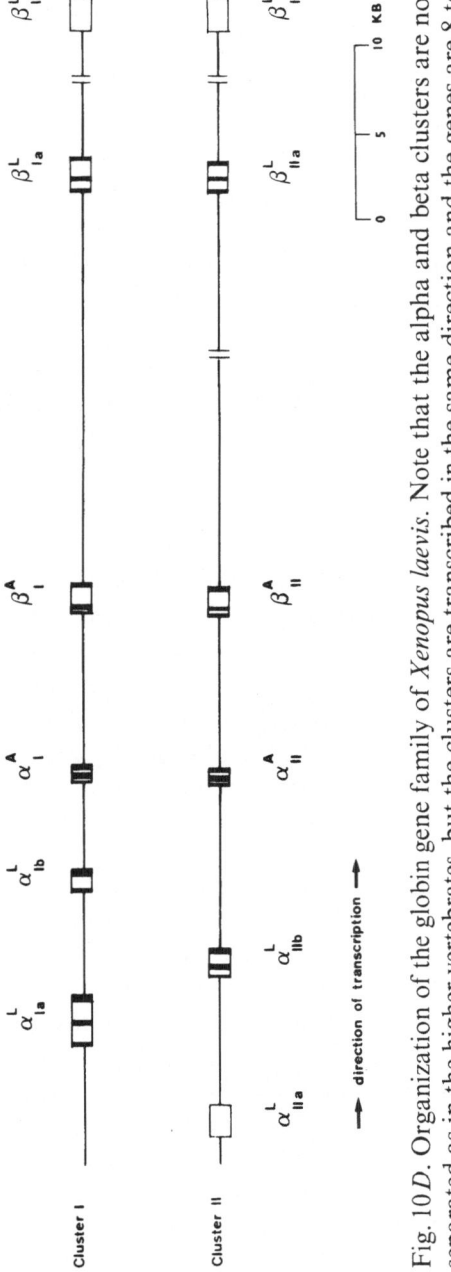

Fig. 10D. Organization of the globin gene family of *Xenopus laevis*. Note that the alpha and beta clusters are not separated as in the higher vertebrates, but the clusters are transcribed in the same direction and the genes are 8 to 10 kb apart. The positions of repeating segments are not yet known. Symbols represent larvae or adult genes, $\alpha^L_{1a}$ and $\alpha^L_{1b}$ code for larval alpha-like hemoglobin subunits in cluster 1; $\beta^L_{1a}$ and $\beta^L_{1b}$ code for larval β-like globins while $\beta^A_1$ is the coding region for the adult β-chains. Cluster II has similar symbols. (From HOSBACH *et al.*, 1983; copyright is held by M.I.T.; photograph provided by Dr. RUDOLF WEBER)

have been detected in the DNA of other vertebrates near the globin genes (VAN DER PLOEG *et al.* 1980), but in no case has any functional correlations been made for these mCpC sites.

*Xenopus laevis* has a different arrangement of the globin genes. The alpha and beta genes are in one cluster with the sequence in the direction of transcription being two larval genes of the alpha type followed by an adult alpha gene and then the single adult beta gene of the cluster. Downstream with a gap of about 25 kb are then two other larval beta genes. Since *Xenopus* appears to have been derived from a tetraploid, or at least to have four sets of globin genes in diploid cells, there is another chromosome with a very similar gene arrangement (Fig. 10 *D*). Studies of the methylation patterns in relation to functional state indicates a good inverse correlation between methylation and function in a particular cell similar to that of human, rabbit, mouse and chicken hemoglobin genes (GERBER-HUBER *et al.* 1983, and personal communication from Dr. G. V. RYFFEL, Bern, Switzerland).


## C. Differentiation of Lymphocytes and the Role of Methylation

Remarkable progress has been made in the last few years in understanding the arrangement of the genes which code for antibodies (immunoglobulins) in mammals. The genetic structures involved are complex and the timing of changes involved in differentiation are not fully understood, but a role for regulation of the gene for the joining protein (J chain) during B cell differentiation has been indicated (YAGI and KOSHLAND 1981). There is also an indication that changes in methylation of the genes coding for the constant regions of the heavy chains occurs at the time of the heavy-chain switch, *i.e.,* rather late in the differentiation of the antibody producing cells. Before trying to understand the possible role of methylation a little background on the operation of the system is probably justified. We will use the mouse and human immunoglobulins, which are similar, as examples. Other animals will be mentioned only briefly for certain comparisons.

The immunoglobulin (Ig) molecules are large proteins found dissolved in blood and other body fluids including lymph, milk and tears. Each molecule of Ig consists of four polypeptide chains which are covalently linked by disulfide bonds and further joined and stabilized into very characteristic shapes by non-covalent forces. The molecules are typically Y-shaped, with two of the four polypeptide chains, being light chains (lower molecular weight) and two being heavy chains. In any molecule the two light chains and the two heavy chains are identical. Each chain is composed of a variable region and a constant region. It is the variable regions which give the molecules their great diversity and potential for reaction with so many different antigens. In a recent review LEDER (1982) has estimated that a species or even one individual has the potential for 18 billion different specific Igs. It is unlikely that the full potential is ever realized or required for the multitude of antigens with which an individual or species comes in contact. In spite of the great diversity, a cell which secretes antibodies (plasma cell) produces only one specific antigenic type. Two of the three arms of the Y-shaped molecules are identical and at certain sites are capable of binding a specific antigen. The third arm of the Y and approximately the inner one-half of the other arms are much less variable from molecule to molecule. The latter sequences and consequent structures are coded by genes which vary no more than other structural genes of the hemopoeitic system. The remarkable differences in the antigen binding regions result from the coding of each chain by two different genes that are on the same chromosome but may be separated by hundreds of kilobases in the early embryonic cells.

The cells which differentiate into lymphocytes, are first seen in the yolk sac along with the other hemopoeitic stem cells. Two types of lymphocytes are recognizable, not by their appearance, but by their behaviour or function. The B cells are those lymphocytes which divide and produce the differentiating cells that mature into plasma cells. One lymphocyte can produce a large clone of cells, some of which retain the properties and specific antibody producing potential of the original cell while others divide and mature into plasma cells. The lymphocyte precursor cells, called pre-B cells, migrate to and reside for a time during development in a sac-like gland called the bursa which buds off the lower intestine in birds; hence the name B cells. Since mammals do hot have a bursa the B-cells differentiate in the embryonic liver. Another group of similar cells migrate to the thymus gland and differentiate into the T cells. These do not give rise to antibody producing cells, but they have molecules resembling Igs on their surfaces and function in the recognition and destruction of certain viral infected cells and other variants including transformed tumorigenic cells, certain parasites and cells of foreign tissue or organ transplants.

Since the only correlations with methylation so far studied relate to B-cell differentiation we will confine our attention to these cells. The arrangement of the antibody genes are shown diagrammatically in Fig. 11 (MARCU and COOPER 1982).

The variable region of each heavy chain is coded by one of 80 or more V genes tandemly arranged upstream from the cluster of constant (C) genes in the order shown. The constant region is coded by one of the eight genes of the downstream cluster. During B-cell differentiation, the variable gene which will be functional is moved to within about 1,500 nucleotides of the mu gene of the constant gene cluster by somatic recombination. The region between is assumed to be deleted by intrachromosomal rearrangements except that two intervening segments are preserved. These are the D and J genes or short sequences that contribute to the variability of the V region of the final Ig molecule. There are a number of D sequences, one of which is preserved along with at least one of the J genes in the multiple recombination events which occur. The mechanism and timing of these

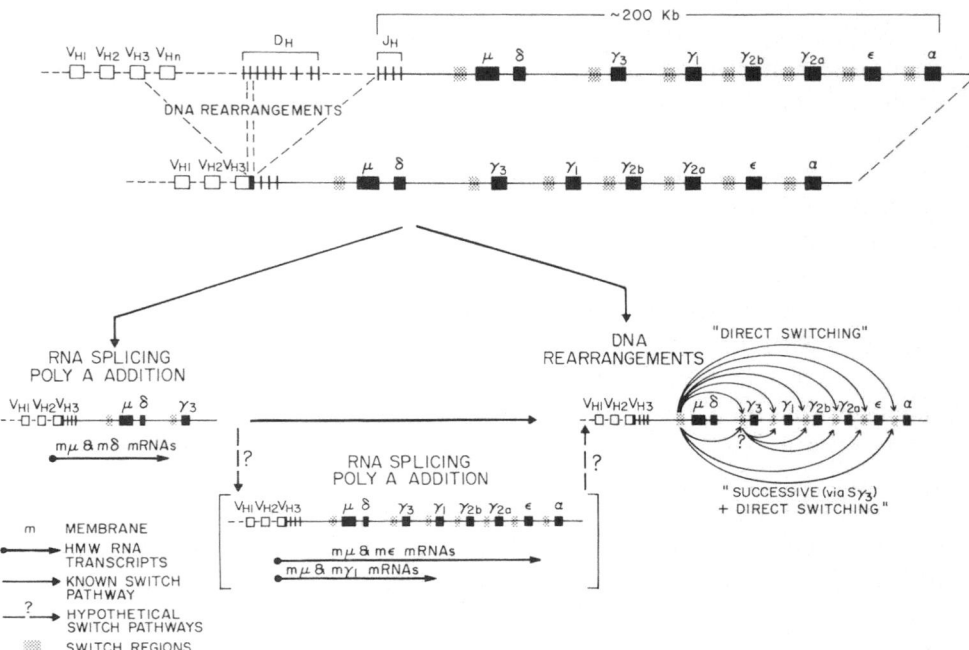

Fig. 11. The arrangement of the mouse immunoglobulin heavy chains in relation to the variable region genes and the switching regions. Alternative ways are shown for connecting variable regions with a specific constant region gene. In the early stages of differentiation the RNA splicing may be utilized, but the switching is the mechanism for most differentiated cells. In the course of B cell ontogeny $V_H$, $D_H$, and $J_H$ DNA segments become assembled to encode a complete heavy chain variable region and later this assembly is associated with different constant segment genes. The first is frequently the $\mu$ gene and it is at this time that the J-gene, which codes for the protein that forms the pentamer molecules of immunoglobulin, is turned on. (From MARCU and COOPER, 1982; reprinted by permission from Nature, Vol. 298, p. 327; photograph provided by Dr. KENNETH B. MARCU)

events is still incompletely understood. It is possible that the deletions involve unequal sister chromatid or homologous chromatid exchanges rather than intrachromatid rearrangements. The net result is known to move one of the V regions along with a D and J region to the vicinity of the mu gene. The sites at which recombination occurs seem to be restricted by the nucleotide arrangement in the switch regions, but in the J region part of the variability of the final heavy chain is contributed by a limited flexibility in the site of recombination in different cells.

The functional light chain genes are assembled by similar recombination events in which a V region and a J region are moved close to the single constant region (kappa or lambda) in any one chromosome. The regions coding for both light and heavy chains contain introns which are removed in the processing of the RNA, but we will not include these complications in our discussion.

The pre-B cells begin producing Ig molecules from the rearranged genes. The heavy chains are produced first but without light chains these neither assemble nor appear on the cell membrane of the immature B lymphocytes. Presumably when the switch has assembled a functional light chain region, the mRNA is produced and Ig molecules begin to appear on the surface of the B cells. The first molecules have a terminal sequence which anchors them in the membrane. This is coded by the last 2 exons of the mu gene and the presence or absence of this anchor sequence is determined by the termination of the primary transcript. There is a stop codon near the end of the 4th exon of the mu gene. If the transcript includes exons 5 and 6, processing removes the stop codon and leaves the anchor as part of the mRNA, but if the transcript stops short of these last two domains the stop codon remains in exon 4 and the resulting molecules are secreted rather than anchored in the membrane.

The mature B lymphocytes then have surface IgM coded by the combination V, D, J and mu regions, but do not yet secrete antibodies. The B lymphocytes can have IgD and probably one of any other class of Igs on the surface, but all antibodies produced from one cell have the same variable region and therefore the same specificity for an antigen. After stimulation by an antigen which reacts with the variable arms of the Ig molecules, the B-lymphocytes begin to divide and the primary transcripts end before exons 5 and 6. The anchor sequence is missing and the antibodies are secreted from the cell. However, these first molecules secreted in an immune response are pentamers, *i.e.,* they are formed of five Y-shaped tetrameric polypeptides joined together by the arm containing the constant region of the immunoglobulins. This pentamer is formed by the interaction with a polypeptide called the J chain. It is coded by a separate gene which is not to be confused with the J region of the Ig variable region. The transcription of the J gene that codes the J chains has been shown to be initiated when secretion begins. The J chains form disulfide bonds with the constant region of heavy chains and produce the pentamers (KOSHLAND 1975).

After producing IgM another switch occurs and the differentiated plasma cell secretes monomers of either IgD, IgG, IgE or IgA, which are coded by the variable region coupled to a delta, gamma, epsilon or alpha gene, respectively. This switch which occurs after stimulation of the lypmphocyte by an antigen that happens to react with the variable region of the IgM molecules on the surface, is a late event during the differentiation of the plasma cells. IgD or one of the other classes of Igs can also be found on the surface of the B lymphocytes, but the mature plasma cell typically secretes only one of the classes of Igs with the same specifity as those its initial pre-B cell had. Mutations can also alter the specificity of the Igs during the divisions which occur during differentiation of plasma cells, but these probably contribute very little to the variability of antibodies.

After stimulation by a specific antigen, a B-lymphocyte divides, and before secretion begins, the J chain gene is turned on. This switch has been studied by MATHER et al. (1981) who also reviewed the earlier work which revealed the mechanism of interaction of J chains with Y-shaped IgM molecules to form pentamers. They also obtained evidence that the switch involves control at the level of mRNA production. By examining several lymphoma cell lines which represent different stages in the differentiation of the lymphocytes, they could test for the presence of J chain mRNA or its precursor. They used a cloned cDNA prepared from J chain mRNA as a probe for the production of J chain RNA in the various cell lines. A lymphoma, WEHI 231, with the properties of a mature B cell, synthesizes monomer IgM but no J chains. When this cell was fused with a plasmacytoma cell, MPC # 11, which secretes IgC, the hybrid cell began secreting pentamer IgM. Neither parental cell secreted IgM or had a detectable amount of the mRNA for J chain synthesis but the hybrid produced both, and the J chains were active in forming pentamer IgM. YAGI and KOSHLAND (1981) showed by comparison of several cell lines, using restriction mapping with readioactive probes, that the change to J chain production does not involve rearrangement of the gene. However, expression of the J chain is correlated with loss of DNA methylation in the J gene region. The deletion of methylation was measured by the comparison of Msp I and Hpa II digests after electrophoretic separation and blotting on nitrocellulose followed by hybridization with the labeled cDNA probe and autoradiography.

The constant genes of the Ig cluster were also probed for methylation patterns by the same technique to determine if the change in the J chain gene was part of a general loss of methylation patterns or a site specific change. The mu constant region was found to be undermethylated in all of the cell lines that produced IgM. All of the cell lines are potential secreters of IgM, but WEHI 231 had not made the switch to one of the other constant genes before being transformed. Only the cell line, MPC # 11, which was a secreter of IgG had the respective gamma constant gene undermethylated. These experiments indicate a positive correlation between loss of methylation at specific regions and function of a gene. However, as with the data from erythrocytes and erythroblasts, the experiments indicate that although undermethylation is a precondition to function, it is not sufficient to initiate transcription. Without other conditions or mechanisms which initiate transcription the unmethylated sites may remain inactive. We may assume that the fusion of cells, which produced an environment necessary for IgM production, also initiated events that resulted in the loss of the methylation pattern at specific sites in the genome, including the region of the J chain gene. The time in differentiation at which the delta, gamma, epsilon or alpha genes lose their methylation is not known, but these observations (YAGI and KOSHLAND 1981) indicate that it is probably late in differentiation when the switch to production of one class of Ig occurs. This switch occurs either by DNA recombination or by a change in processing of the transcript into a functional mRNA (MARCU and COOPER 1982).

## D. Suppression of Integrated Viral Genomes in Cells by DNA Methylation

### 1. Small DNA Viruses

Viral genomes of most replicating viruses in mammalian cells are not detectably methylated in their CpG sites (DIALA and HOFFMAN 1982). For example, polyoma has eight Hpa II sites and SV 40 has one, but these are not methylated in the host cell (FORD et al. 1980) nor in the mature viral particles. Likewise Adenovirus type 12 has been examined and found to be unmethylated in the Hpa II sites (SUTTER and DOERFLER 1980). Papilloma virus type 1a has been reported to have about 40% of one of the four Hpa II sites methylated (DANOS et al. 1980). On the other hand integrated viral genomes are frequently methylated and the extent to which they are expressed appears to be an inverse function of the extent of methylation.

DESROSIERS et al. (1979) examined several lymphoid cell lines established from tumors induced in squirrel monkeys by *Herpesvirus saimiri*. Two cell lines which do not produce detectable virus have the integrated viral sequences highly methylated at CpG sites that can be probed by digestion with Msp I and Hpa II followed by electrophoresis, blotting and autoradiography of the DNA that hybridizes with a $^{32}$P-labeled viral DNA probe. On the other hand, three lymphoid lines which are virus producers were not detectably methylated in the viral sequences of the cellular DNA. Most of the viral sequences are integrated into the cellular DNA but in one cell line 1670, a non-producer, part of the viral sequence is present as an episomal DNA, a closed circular form. That episomal DNA, as well as the CpG sites in the integrated sequences, was found to be highly methylated and no infective virus was produced by this cell line.

SUTTER and DOERFLER (1980) examined free adenoviral DNA isolated from purified virions and found no significant methylation at Hpa II sites. On the other hand, four lines of adenovirus (Ad 12) transformed hamster cells had the integrated genomes extensively methylated. The segments of integrated viral DNA that code for early proteins, which were expressed as mRNA are undermethylated compared to the genes coding for late proteins in two transformed cell lines. In contrast, in two lines of Ad 12-induced rat brain tumor cells, some of the late genes were expressed. In these cell lines the DNA which comprises the late genes is undermethylated compared to the extensively methylated regions in the Ad-12 transformed hamster cells. In each instance there is an inverse correlation between levels of methylation and expression of the integrated viral genes.

YOUSSOUFIAN *et al.* (1982) studied the methylation of herpes simplex virus (HSV) in relation to latency. The virus usually becomes latent in vivo after a primary infection, but the study of the viral DNA in vitro is difficult. They used a model system developed by HAMMER *et al.* (1981) to study the methylation of the DNA in the latent stage. A lymphoblastoid T-cell line, CEM, is persistently infected by HSV-1. However, occasional latent periods were found when the cells were propagated for long times. Cells in latent stages harbor only one or two copies of the viral genome whereas the producing cells have 40–80 copies. During periods when cells were producing viruses the viral DNA was not detectably methylated, but about 800 days after the initial infection when the productive cells became latent or non-productive of virus, the few remaining copies per cell were heavily methylated in the Hpa II sites. Most or all of the viral genome is present in cells in the latent stages and no differences in sequence arrangement could be detected from analysis by restriction endonuclease digestion, electrophoresis and blot hybridization.

## 2. Inactivation of Retroviral Genomes by DNA Methylation

Retroviral genomes can be introduced into cellular genomes at various stages of development or they can be inherited as a Mendelian gene in the germline (COHEN 1980, GROUDINE *et al.* 1981, HARBERS *et al.* 1981). The viral genomes are methylated to various extents and the expression of the viral genomes is suppressed when the genomes are highly methylated.

Mouse mammary tumor virus (MMTV), which is responsible for mammary carcinomas in many inbred strains, has been studied as a model system for the influence of methylation on expression of genes (COHEN 1980). By the use of the restriction endonucleases Msp I and Hpa II, as well as Hha I, which cleaves the unmethylated 5'-GCGC sites, he showed that viral genomes acquired by an individual through the germ line were extensively methylated. On the other hand, the viral genomes integrated after birth by milk borne infection were not modified at the same sites. DNA from normal infected tissues or transformed tissues showed various degrees of hypomethylation of the viral genomes.

GROUDINE *et al.* (1981) studied the methylation of a genetically transmitted endogenous virus in chickens. Most chickens contain in their DNA at least one genetic locus, *ev-1*, which can be identified in "Southern" blots as a 10.5 kb Sst I restriction fragment. Among 10 cell lines with genetic loci which are similar in structure, ASTRIN 1(978) has described one, *ev-3*, which contains a portion of the typical viral genome integrated in a way that serves as a template for 31 S and 22 S viral RNA transcripts. However, these cells carrying an *ev-3* locus do not produce either infectious virus or viral particles as do other cells carrying the integrated genome. The *ev-3* cells contain a 120 kilodalton polyprotein, which is believed to represent a non-functional form of the precursor to reverse transcriptase, a protein which is regularly found in cells infected with a retrovirus and presumably is coded by the *ev-3* locus. On the other hand, cells carrying the methylated *ev-1* structural sequence, maintained it without detectable expression. When the methylation of the Hpa II sites were compared in these two regions of the genome, the *ev-1* sequence was found to be not only methylated, but also relatively resistant to DNase I digestion, while in *ev-3* the similar sequence was undermethylated and sensitive to DNase I at certain sites. When cells of *ev-1* were grown for a short time in a culture medium containing azacytidine, the expression of the viral genome was induced. Analysis of cytoplasmic RNA showed sequences that hybridized with $^{32}P$ labeled cDNA containing viral sequences that hybridized with $^{32}P$ labeled cDNA containing viral sequences. At least one DNase I sensitive site was acquired in the viral sequences of *ev-1* as a result of the change induced by azacytidine. Azacytidine, incorporated into DNA after conversion to a precursor molecule, inhibits DNA methylase and allows some sequences to replicate with the consequent loss of their methylation patterns.

One notable exception to the observations that productively infecting viral DNAs are not methylated has been reported by WILLIS and GRANOFF (1980). The DNA extracted from the frog virus 3 (FV 3) virions is highly methylated. More than 20% of the cytosines are methylated. The virus can also be grown in cultured cells of the fathead minnow and in BHK 21 cells (baby hamster kidney) which methylate their DNA to lesser extents, 8% and 2% respectively. It is possible that the frog viral genome codes for a methylase since a methylase has been reported to be induced in infected cells. The methylase was found predominantly in the cytoplasm. FV 3 DNA in the nucleus of the host cells was not methylated, whereas that in the cytoplasm was methylated. The role of the methylation and its effect, if any, on transcription is unknown.

### E. The Vitellogenin Genes in *Xenopus* Are Methylated, But Can Be Expressed Without a Detectable Change in the Pattern

Another exception to the long list of instances in which there is a correlation between methylation and suppression of gene expression has been revealed by studies of the vitellogenin genes in the frog, *Xenopus*. The functional vitellogenin genes are in the frog liver cells. These genes have been mapped and their expression studied in relation to hormonal induction (WAHLI *et al.* 1980, WAHLI *et al.* 1981). The genes are expressed in egg-producing females but not in males. However, synthesis can be induced in the liver of males by injection of estrogen. Large amounts of the protein then accumulate in the blood. The genes are present in single copies, but there are two related genes, designated A 1 and A 2. The A 1 gene is 21 kb and A 2 is 16 kb, but both are interrupted by 33 introns of various sizes. The average intron length for A 1 is 0.45 kb and for A 2 is 0.31 kb. The dispersed repeated sequences in and flanking the gene have also been mapped (RYFFEL *et al.* 1981). Six different repetitive sequences within the transcribed introns of the A 1 gene have been mapped and shown to be transcribed in non-vitellogenic liver cells. The presence of transcripts in these cells suggests to the authors that these sequences of middle repetitive DNA are present in transcriptional units which are active in many cells in the absence of estrogen.

The vitellogenin genes have a number of Hpa II sites which are methylated at the CpG cytosine. The pattern is the same in males and females and in various tissues such as sperm, early embryo, liver and other adult tissues. There is no change in the methylation pattern when the genes in the liver of males are induced to secrete the yolk proteins by injection of hormone. Of course, not all of the CpG sites can be monitored by the Hpa II-Msp I technique, but any deletion or dilution of all methylcytosines in the functional genes can be ruled out (GERBER-HUBER *et al.* 1983).

DNase I digestion of the chromatin in various cells was compared on the basis of a mathematical model and expressed by a sensitivity factor. The sensitivity varied over a wide range and was correlated with activity of the genes analyzed. The beta one-globin gene fragment was more sensitive in erythrocytes than in hepatocytes, whereas the albumin gene was more sensitive in hepatocytes. However, the A 1 vitellogenin gene had the same sensitivity in both types of cells. All three genes were inactive in the cells examined but have different potentials for stimulation (FELBER *et al.* 1981). For example, when the same type of comparisons were made for liver cells from males treated with estrogen, the entire A 1 and A 2 vitellogenin genes were about twofold more sensitive to DNase I in chromatin of hepatocytes from treated males than untreated ones. Analysis of the DNase I sensitivity of the other two genes, beta one-globin and albumin, demonstrated that the genes remained unaltered by the estrogen (GERBER-HUBER *et al.* 1981).

NOWOCK and SIPPIL (1982) have detected a class of proteins in nuclear extracts which bind to dispersed repeats in the 22.2 kb segment including the chicken lysozyme gene. Four restriction fragments with sites flanking the transcribed genes bind the proteins competitively. Two sites are 6.1 and 3.9 kb upstream from the transcription start site. Two others are 2.8 and

6.2 kb downstream of the poly(A)-addition point. The binding constant for one of the DNA fragments was estimated to be $6 \times 10\text{–}12\,M$ and since the fragments are mutually efficient competitors, one class of proteins is involved.

Sequencing one of these regions (SIPPEL, personal communication) showed a dispersed repeat upstream from one of the receptor protein binding sites that is represented by many copies in the genome. The binding site has some sequences in common with the dispersed repeat but these were insufficient to hybridize detectably with the same class of dispersed repeats. The interpretation of these relationships is not yet possible, but the possible role of dispersed repeats in the control of gene function is an intriguing problem with few clues so far to reveal how they might operate.

## F. Integrated Retroviruses Are Methylated Early in Development

Recently Jaenisch's group in Germany (JAHNER et al. 1982) reported that methylation of the proviral genome of retrovirus occurs early in embryonic development. The methylated provirus is suppressed in the expression of its genes and is ineffective in transfection assays. HARBERS et al. (1981) studied some substrains of mice which carry the integrated Moloney murine leukemia virus (M-MuLV). One substrain, Mov-3, carries the provirus in a form that is activated during embryogenesis and the embryos developed localized viremia and finally the animals develop leukemia at some age after birth. This provirus was cloned into pBR 322 as a 16.8 kb Eco RI fragment. Comparison of the cloned and genomic sequence by restriction enzyme analysis showed no differences. On the basis of this analysis they assumed that no major sequence rearrangements occurred during cloning.

When the genomic and cloned DNA was compared for infectivity, the cloned DNA was found to be more than 100 times more effective than the genomic DNA. The only detectable difference was that the cloned DNA had lost the methylation pattern at CpG sites as indicated by digestion with Hha I. JAHNER et al. (1982) then conducted experiments to find out when in embryonic development the provirus was methylated. Microinjection of zygotes with virus or infection at pre-implantation stages (4–16 cell embryos) results in integration of multiple copies of the viral genome into the mouse chromosomes. However, the DNA was not active in transfection assays when extracted from 2–4 months old animals. In addition, there was no evidence that the viral genes were expressed in the embryos of the mice infected in early development. All cells of mice infected at the zygote stage had the same copy number of integrated proviruses in all of their cells and at the same chromosomal sites. Those infected at the 4–16 cell stage were mosaics with various numbers of copies at a few different locations in the chromosomes from the various tissues. These data indicate the M-MuLV genomes were integrated soon after infection and little viral production or reintegration occurred during subsequent stages of development. On the other hand, if the embryos were infected at day 8 of gestation when the embryos consisted to $10^4$–$10^5$ cells, integration occurred, but the pattern was very different from that observed in earlier infections. The viruses had integrated about the same number of copies per mouse genome, but at many different sites in the various cells of liver, kidney and brain. This pattern indicates foci of viral production, reinfection and spread in prenatal embryos.

When the methylation patterns were examined by the use of Hpa II, Hha I and Ava I digestions, all of which have CpG sites in their recognition sequence and are unable to cleave the DNA when the cytosine is methylated, the proviruses integrated before implantation were found to be highly methylated. By contrast those integrated by infection at day 8 of gestation were not methylated. Some foci of infection occurred in adult mice derived from both classes of embryos, but the DNA could be distinguished because proviral DNA integrated from the early infection was inactive in transfection. On the other hand, the unmethylated DNA from the cells of animals infected late in embryonic development was active in transfection even when no free virus could be detected. The DNA from animals infected early in development could be made active in transfection by cloning and amplification in bacteria where the methylation pattern would be lost.

The experiments by MINTZ's group (STEWART and MINTZ 1981) suggest a similar situation, in that the transformed euploid teratocarcinoma cells lose their tumorigenic properties when injected into mouse blastocysts. The injected cells participate in the formation of all tissues including the gonads which allows transmission to the next generation. The genes producing the tumorigenic transformation in the teratocarcinoma cells are unknown and the methylation if it occurs can not be followed at the relevant genomic sites. It is likely, however, that a similar de novo methylation suppresses the oncogenes in teratocarcinoma cells.

## G. Suppression of Metallothionein Genes by Methylation

Metallothioneins (MTs) are small, cysteine-rich proteins that bind heavy metals such as Cd, Zu, Cu, Ag and Hg. These proteins are intracellular and widely distributed in both vertebrate and invertebrate tissues. They appear to be involved in zinc homeostasis and detoxification of heavy metals. The synthesis of MTs can be regulated by heavy metals and by glucocorticoid hormones (KAGI and NORDBERG 1979, and KARIN et al. 1980). To study the regulation of these proteins BEACH and PALMITER (1981) selected strains of Friend leukemia cells resistant to cadmium. These cells could be induced to synthesize metallothionein to the extent that 70% of the incorporated cysteine was bound in these molecules. The synthesis was inducible with $25 \mu M$ $CdSO_4$ for 8 hours followed by a medium with $145 \mu M$ $CdSO_4$ and/or $80 \mu M$ $ZnSO_4$. They showed that the synthesis was under transcriptional control by measuring the MT mRNA in cadmium resistant and sensitive cells. Examination of rates of transcription in isolated nuclei of metallothionein resistant and sensitive cells indicated that the resistant ones transcribed the sequences at about 3 times the rate for sensitive cells. When the number of copies of MT genes was measured by hybridization with a labeled MT sequence probe, the results indicated the resistant cells had amplified the genes to about threefold which could account for the increased rate of transcription. The cells could be induced to synthesize even more metallothionein proteins by either exposure to the appropriate concentration of cadmium or by glucocorticoids in the medium.

COMPERE and PALMITER (1981) examined a mouse thyoma cell line, W 7, which is not inducible with either cadmium or glucocorticoids. In this cell line all of the Hpa II sites in the vicinity of the MT gene were methylated in contrast to other cell lines which are inducible. The latter have all of the Hpa II sites unmethylated. Treatment of the W 7 cells by growth in medium with $5 \mu M$ azacytidine for 12 hours followed by transfer to fresh medium with $10 \mu M$ cadmium increased their concentration of MT mRNA six- to eightfold. However, if DNA synthesis was prevented during exposure to azacytidine no change in inducibility could be detected. Some of the inducible cells were grown for approximately 60 cell doublings in the presence of $10 \mu M$ cadmium during which cells able to produce metallothionein should have had a selective advantage. The cells were then challenged with $60 \mu M$ cadmium. Control cells, which had not been grown in azacytidine and followed by selection in $10 \mu M$ cadmium, had no detectable MT mRNA when challenged, but the selected strain produced 880 MT mRNA molecules per cell.

## H. Other Genes That Are Suppressed by Methylation

There are now a number of other reports of correlations between hypermethylation and suppression of genic expression, and the list grows each month. For example, MANDEL and CHAMBON (1979 a, b) reported on the distribution of methylated and unmethylated Hha I and Hpa II sites in and around the genes coding for ovalbumin, conalbumin and ovomucoid in the chicken oviduct. In general the sites in tissues where the genes are being expressed are less methylated than sites where expression is not occurring, but there are a number of exceptions. Even sperm DNA, which usually has the highest level of methylation has some sites in the DNA of the regions probed that are always unmethylated. On the other hand, some sites were resistant in all tissues tested, and a number of allelic variants for methylation of Hha I and Hpa II sites in the region of the ovalbumin gene were found.

## I. Correlation Between DNase I Sensitivity of Chromatin and Cytosine Methylation

Fig 12 shows a map of the ovalbumin gene and the position of restriction sites of four endonucleases used in a study by MANDEL and CHAMBON (1979 a, b). There are 5 Hha I sites, 3 Hpa II sites, 3 Kpn I sites and 6 Eco RI sites. Since the Hha I sites and the Hpa II sites contain CpG doublets which are the methylatable sites, these two endonucleases were used as probes for methylation and the other two were used to produce gene segments of a size which could be separated by electrophoresis.

Five Hha I sites will be referred to as Hha 1, Hha 2, etc. Hha 1 was not detected as a cleavage site in these studies either because it is always methylated or because in the population studied, this site was not present. Considerable heterogeneity for some of the sites was found in the population of animals used, but these were the ones that were always methylated or unmethylated and presumably played no role in control of the genes in question. No isoschizomer is available to probe for the presence of the Hha I site, as can be done for the Hpa II site with Msp I.

Three classes of sites were identified. One designated $m^-$ are sites containing a CpG doublet in which the cytosine was never detected in the methylated state. A second class designated $m^+$ was methylated in all tissues examined. A third class which is the one of greatest interest was designated $m^v$. The $m^v$ sites varied in methylation in different tissues and may be important in regulatory phenomena.

Site Hha 2 is located in exon 1 of one cloned ovalbumin gene, but there are allelic variants which do not have this site. Analysis of the DNA of 16 laying hens revealed only partial digestion in 5 animals (a mean value of 30%). The other 11 were completely protected at the site in most tissues; in liver DNA there was less than 4% digestion, and digestion was barely detectable in erythrocyte, kidney, spleen and brain DNA.

Sites Hha 3, 4 and 5 mapped in a cloned 9.2 kb fragment "a" (Fig. 12). Site Hha 3 is located about 1,200 bp downstream from the sequence coding for the poly A addition site of ovalbumin mRNA. Hha sites 4 and 5 are further downstream roughly 1,500 and 3,500 kb respectively. A cut at Hha 3 divides the 9.2 kb fragment into two fragments, 3.2 and 6.0 kb long. Only the 3.2 kb fragment hybridized with the ovalbumin C DNA probe used. Digestion of ovalbumin DNA cuts about 85% at this site compared with only 30% in erthrocyte DNA.

For both Hha 2 and 3 no digestion of sperm DNA could be detected even though other tissues were partially digestable, 50% for liver, and less for other tissues. Hha 4 and 5 were never digested to a detectable extent in any of the 15 animals tested; presumably these sites were always methylated.

To summarize, only the Hha site 3 belongs to the $m^v$ group. An investigation of the resistance of the DNAs with methylated and unmethylated sites to DNase I showed that the DNA with an unmethylated cytosine at site Hha 3 was cut much more than the intact 9.2 kb fragment which was methylated and protected from Hha I. About 80–85% of the DNA in oviduct was reduced to fragments of 6.0 and 3.2 kb. The fraction

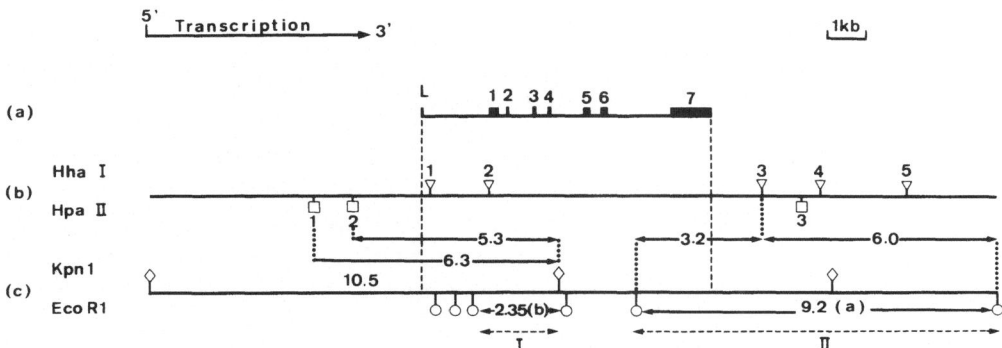

Fig. 12. Map of chicken DNA in the ovalbumin gene region. *a* Location in the ovalbumin gene of the leader-coding sequence (*L*) and the seven other exons. *b* The location of Hpa I and Hpa II sites in and flanking the ovalbumin gene. *c* The Kpn I and the Eco RI restriction sites. The sequences for the probes I and II used in the analysis are shown by horizontal dashed lines. (From MANDEL and CHAMBON, 1979 b)

resistant to DNAse I was 15–20% of the total DNA and corresponds to the percentage of non-tubular gland cells in the oviduct. Since only the tubular gland cells secrete ovalbumin, the DNA resistant to Hha I is likely to be the DNA from cells not participating in the transcription of ovalbumin gene (KUO *et al.* 1979).

## J. The Hpa II Sites in Ovalbumin Gene

Three of these sites are present; two are located in the 10.5 kb segment released by Kpn I and the other is near the middle of the 9.2 kb unit released by Eco RI (Fig. 12). In DNA from a laying hen, the 10.5 kb fragment when digested by Msp I was reduced to a 5.3 kb fragment which still hybridized with the probe; the other end of the fragment was 5′ to the region of the probe. This proved that the site was present in the homozygous state, but only 8% digestion occurred in erythrocyte, liver and kidney DNA from the same animal. Oviduct DNA was converted to the shorter fragment (40%) which indicates unmethylated sites in the functional DNA even if the amount was lower than expected if both chromosomes were active. Two other laying hens had some of the ovaduct DNA which was digestable by Hpa II, but the Hha 1 and 2 sites appeared to be quite variable, probably because these sites were absent or heterozygous in some of the animals. The Hha 3 site was also difficult to analyze. Msp I did not digest the DNA to completion, perhaps because some of the sites had the outer cytosine methylated at the recognition site.

## K. The 5' End of the Rat Albumin Gene Is Undermethylated in Cells in Which It Is Expressed

OTT et al. (1982) examined clones of rat hepatoma cells that vary quantitatively in their rates of synthesis of albumin. These were compared along with other variant and hybrid cells which produce no albumin. The clones are of four classes: 1. well differentiated cells that express several hepatic functions as well as produce albumin, 2. dedifferentiated variant cells that fail to express the hepatic functions including serum albumin production, 3. albumin producing revertants derived from the above variants which procedure albumin and 4. somatic hybrids formed by fusion of variants with the well differentiated parental cells. The number of copies of mRNA per cell was determined and shown to be proportional to the amount of albumin secreted per cell. The gene segment analyzed by OTT et al. has ten Eco RI sites and five Hpa II sites, which indicates that the gene has probably lost a number of the original CpG sites by mutation to TpG as discussed later. The methylation of four of these Hpa II sites was studied by the Southern blotting techniques described above in order to determine the percentage of methylation at each site. These percentages were compared with DNA from the liver of ACI strain male rats, from which the original hepatoma clones were isolated.

In the normal liver cells the sites were methylated to the extent of 10 to 20%. Since the liver is composed of 80% hepatocytes and all of these synthesize albumin, these data are consistent with all functional genes being unmethylated at the sites examined. Three clones that have stable differentiated phenotypes, in which all cells produce albumin, have no methylation at Hpa II site 1, whereas the three other sites tested have 95–100% methylation. One clone Fu 5-5 has about 25% methylation at Hpa II site 1, and 15–20% of the cells failed to show albumin synthesis when assayed by an immunofluorescence technique. Two subclones of this clone were examined. One, c 16, has no methylation of site 1 but 100% methylation at the other sites. All cells in this clone synthesize albumin.

Another subclone c 17, with 60% of the Hpa II site 1 methylated, had all cells weak or negative in the microscopic immunofluorescence test for albumin. This is excellent correlation with function and non-methylation of the albumin gene at site 1. Methylation of the other sites does not affect the function of the gene.

Variant cells which had lost expression of the liver functions, including the synthesis of albumin, were highly methylated at site 1, but the level varied from 50% methylation in one clone to 85–100% in several other clones. Yet all failed to produce albumin.

Tests on revertants gave mixed results. C 2 cells, one of the derivatives from a differentiated cell line, FAU, gave rise to cells of altered morphology which produce significant amounts of albumin. These clones had from 0 to 99% of site 1 methylated. Some of the lines with as much as 80% methylated had 40–50% producers. Here the correlation was less than expected, if methylation at site 1 effectively prevents production of albumin.

In spite of the excellent correlation between function and under-methylation of the 5' end of the albumin gene the authors argue against the possibility that the methylated state of such genes determines whether or not they will be expressed during development. Why they take this point of view is not clear from their discussion; certainly others looking at the data could conclude that methylation around the 5' end of a gene can be incompatible with its transcription. Certainly one would agree that loss of the methylation pattern at a CpG site will not turn on a gene, but it could alter the affinity for regulatory molecules, exerting either a negative or positive control, if these were available in the cell.

## L. Methylation of Human Growth Hormone and Somatotropin Genes

HJELLE et al. (1982) reported an inverse correlation between methylation and expression of the genes which code for human growth hormone and chorionic somatomammotropin. They compared the Hpa II and Hha I sites in and around the genes in pituitary gland and placenta, where the genes are expressed, with corresponding sites in cells from other tissues, leukocytes and a hydatiform mole, where expression does not occur. The extent of methylation at these sites was greater in leukocytes with decreasing amounts in pituitary gland and placenta. The methylation of DNA in cells of the hydatiform mole was about equal to that in placenta. Here again, we see a correlation between expression and undermethylation, but the under-methylation is not a sufficient condition for expression of these genes. There is little or no expression of the genes of the hydatiform mole, but the DNA of the two genes has the same methylation level as the placental DNA in which expression does occur. In addition, the level of methylation in the pituitary and placenta is not correlated with the extent of gene expression.

## M. Methylation of Ribosomal Genes

There are also several reports of correlations between the level of methylation and expression of ribosomal genes. BIRD *et al.* (1981 a) examined the methylation of mouse ribosomal DNA (rDNA) using restriction endonucleases. Most rDNA can be cleaved by these enzymes, that can not cleave DNA sequences in which the CpG sites are methylated, but all tissues have a fraction that is resistant. The size of the methylated fraction varies slightly between individuals of the same inbred strain, but much less than between individuals of different strains. The heterogeneity of spacer region enabled BIRD *et al.* to demonstrate that each nucleus contains methylated and unmethylated rDNA. Only the unmethylated copies were hypersensitive to DNase I. They proposed that the methylated genes are the inactive ones. BIRD *et al.* (1981 b) showed that the rDNA of *Xenopus* blood cells is heavily methylated, but two regions in the spacer sequences are frequently unmethylated. These unmethylated regions coincides with two regions containing a 60 nucleotide tandemly repeated sequence which is present in all somatic tissues tested. Sperm rDNA by contrast is methylated in these sites. Loss of this methylation pattern occurs progressively over the first 20 hours of cleavage in the early embryo, during which period rRNA synthesis is initiated and increases in rate.

MACLEOD and BIRD (1982) studied the expression of the ribosomal genes in hybrids between *Xenopus laevis* and *X. borealis*. They found that 97–98% of rRNA precursor in hybrid tadpoles was of the *X. laevis* type, although the *X. laevis* and *X. borealis* rDNA was equivalent in the level of methylation. However, they found that the *X. laevis* rDNA was hypersensitive to DNase I. The study shows that hypomethylation, and even DNase I sensitivity of the *X. borealis* rDNA, was insufficient to ensure its transcription. Examination of sperm rDNA of *X. borealis* showed that hypomethylated sites are located in the same spacer locations as in somatic cells. This contrasts with *X. laevis* where hypomethylated sites are detectable in the spacer of somatic rDNA, but not in the sperm rDNA. Moreover, the loss of spacer methylation, that occurs in early development in *X. laevis,* does not occur in *X. borealis*.

TANTRAVAHI *et al.* (1981) reported that amplified rDNA in rat hepatoma cells is methylated and inactive. In this cell line, H 4-11 E-C 3, there is a tenfold excess of rRNA genes. Two assays, antibodies to 5-methylcytosine and digestion with Hpa II and Msp I, indicated hypermethylation of amplified DNA. Activity of the ribosomal DNA was indicated by staining with silver in a technique that stains active nucleoli. On the basis of this test most of the amplified regions were inactive. Similar studies of a locus on chromosome 14 of the human complement indicated an amplified region of rRNA coding DNA. The DNA was transcriptionally inactive and highly methylated. Recently KUNNATH and LOCKER (1982) examined the methylation level in various tissues in relation to development and growth in the rat. At 14 days gestation the rRNA genes of the liver were mostly unmethylated, but by the 18th day 30% were methylated. At day 18 the rDNA in liver was rather uniformly methylated, when methylated at all, but in adult tissues the pattern was discontinuous with unmethylated genes intercalated in methylated regions. In a tissue culture line the continuous to discontinuous pattern could be induced by transformation of cells with Kirsten sarcoma virus. In adult tissues the lowest level of methylation was in rapidly growing cells of the jejunal epithelium and the highest level was in spermatozoa. These studies indicate a correlation of methylation with inactive rDNA.

# IV. DNA Methylation and the Inactive X Chromosome of Mammals

## A. A Brief History of X Chromosome Inactivation Studies

In 1949 BARR and BERTRAM (1949) reported that the neurons of female cats contained a peculiar body of heterochromatin adjacent to the nuclear membrane which was absent in cells of the males. Later studies showed that many somatic cells of female mammals had this body. In 1960 TAYLOR using $^3$H-thymidine and autoradiography reported that the two X chromosomes of Chinese hamster females were different in the pattern of DNA replication. The long arm of one X, which is heterochromatic, is replicated late in S phase while the short euchromatic arm and the whole of the other X chromosome is replicated in the first half of the S phase. The single X in a culture of male somatic cells is replicated in the first half of S phase, similar to the early replicating X in female cells. At about the same time LYON (1961), who had been studying the coat color variegation by a gene translocated from an autosome to the X chromosome in mice, explained her results on the hypothesis that one X chromosome is inactivated and that the heterochromatic Barr body represented that chromosome. She was unaware at that time of the difference in patterns of replication, but GERMAN (1962) who knew about the work on replication reported that one human X chromosome was late replicating and suggested the connection to the Barr body. Additional autoradiographic studies of replication patterns and karyotypes from humans with aberrant X or Y chromosomes showed that all but one X is late replicating in either females or males with two or more X chromosomes. Extra X chromosomes are heterochromatic and probably genetically inactive (GRUMBACH et al. 1963). The autoradiographic method of identifying the late replicating X following a pulse label with $^3$H-thymidine made identification of the presumed inactive X easy in a variety of species. The hypothesis of an inactive X was verified by cloning cells from a human heterozygous for two alleles of the G 6 PD gene which produced electrophoretically different enzyme molecules. Since G 6 PD was known to be sex-linked, DAVIDSON et al. (1963) analyzed 10 selected clones to see how many of these had one electrophoretic type of the enzyme. All clones carried either one or the other, but not both, of the electrophoretic variants of the enzyme. Later studies have verified this inactivation, but also revealed a segment of the human X chromosome which is not inactivated. The genes, *Xga* (identified by its antigens) and *STS* (steroid sulfatase) were shown to be on the end of the short arm and to remain active when the remainder of the X chromosome is inactivated (reviewed by GORDON and RUDDLE 1981). Two other sex-linked genes HPRT (hypoxanthine phosphoribosyl transferase) and PGK (phosphoglycerate kinase) have also been shown to be inactivated on the extra X chromosome in somatic cells.

Both X chromosomes are active in the preimplantation XX embryos and random inactivation of one X occurs in the fetal tissues after implantation; see KRATZER and CHAPMAN (1981) for a review of the evidence. Once inactivated in a particular somatic cell, the descendents of that cell maintain the same inactive X chromosome for many cell cycles which accounts for clones that produce only one electrophoretic variant of G6PD, for example. In this way one dose of X-chromosome linked genes in females is balanced with two autosomal genes as they would be in cells of the XY males. Reversions to the active state occur at a rate of about $10^{-6}$ per cell generation (MOHANDAS *et al.* 1981), which is comparable to or only one magnitude more frequent than the mutation rate for many genes in mammalian cells.

The genetic activity of the X chromosomes in the germ-line cells is different from somatic cells. As one would have predicted from the abnormal phenotypes of X0 individuals some cells including the oocytes require two active X chromosomes for a normal function. This situation could arise by a failure of inactivation in precursor cells of the germ line or to reactivation of a previously inactive X chromosome. Primordial germ cells have been reported to have a Barr body in some animals but not in others, and biochemical studies have yielded evidence that both X chromosomes are active in germ cells of human embryos at 13 weeks, but not at 12 weeks, of gestation. Thus, reactivation of the inactive X chromosome occurs in oocytes.

KRATZER and CHAPMAN (1981) used an electrophoretic polymorphism at the G6PD locus to study the active genes in X chromosomes of the mouse, *Mus caroli*. G6PD is a dimeric enzyme molecule and if both genes are active in the same heterozygous cell, three molecular species can be produced. In examining electrophoretic gels, made from crushed cells assayed for the enzyme by a technique that allows one to detect the catalytic product of the enzyme, they could identify bands representing the positions of the three dimeric species of the enzyme. At day 10 of gestation no heterodimer band can be detected, but beginning on the 11th day the amount of heterodimer increase gradually until at day 14 the pattern is the same as in mature oocytes. At that time the amount of heterodimer is about 2 times that of either homodimer. These observations indicate that one gene is inactive up to the 11th day and then the gene on the inactive X is turned on again when the whole X chromosome is presumably reactivated.

The number of germ cells increases 2.8-fold between day 10 and 11 and also between days 11 and 12. After day 12 the number remains constant at 11–12 thousand per ovary. Therefore, cell division and DNA replication are proceeding during the reactivation. No attempt was made to determine if all cells of the ovary contained two active X chromosomes, but all cells which differentiate into oocytes apparently have two active X chromosomes. However, X0 female mice (*Mus musculus*) are fertile and the oocytes have only one X chromosome (CATTANACH 1962). X0 females have a shortened fertility life-span and a reduced activity of X-chromosome coded enzymes in oocytes when compared with XX females. In addition the eggs developed from the oocytes of X0 females manifest impaired development during cleavage. This indicates that, even in mice, contributions from genes on both X's are required for normal oogenesis. In human X0 females the oocytes degenerate or the gonical cells fail to differentiate into oocytes. Such females are infertile and have streak gonads devoid of germ cells.

The control of X-chromosome activation or inactivation has been a matter of speculation since its discovery. The first model proposed (GRUMBACH et al. 1963) assumed an episomal factor was released in the egg or cells in the cleavage stages when inactivation occurs. It was assumed to have a specific site on the X chromosome for integration and once integrated it would produce an inhibitor so that no further integration could occur in the same cell. This would account for the random activation of only one chromosome, even if there are one, two, three or more X chromosomes. The spread of the effect from a single site to other parts of a rather large chromosome was not explained without added assumptions.

Another hypothesis which is similar to the above, in that it assumes an inactivation center was proposed by OHNO (1969, 1972) and LYON (1961, 1971, 1972). However, the assumption was made that new protein molecules (DNA binding proteins) were synthesized and were cooperatively bound to the inactivation center on an X chromosome. Because the molecules are limited in number the first site to bind the molecules would quickly use up the supply and additional chromosome would remain inactive. It was assumed that later a generalized repressor substance would inactivate any chromosome not having the inactivation center protected.

A third model was proposed by RIGGS (1975), who likewise assumed an inactivation center which would be methylated by a slow acting methylase similar to the type I bacterial methylases. Since the first site methylated would then increase the chance for the other DNA strand to be inactivated by a methylase that operated hundredfold faster on hemimethylated DNA than on unmethylated DNA, the chance of more than one X chromosome being affected would be small. The methylation pattern once established would be inherited by rapid methylation of the hemimethylated DNA after each replication by the same enzyme which initiated the change. However, Riggs assumed another change to protect other X chromosomes from activation, a new protein to modify the methylase so that it would no longer methylate the unmethylated inactivation centers. A second protein would condense and inactive the other X chromosomes by an unspecified mechanism.

All of these proposals have one thing in common, *i.e.*, the assumption of a single inactivation center on each X chromosome. This appears to be necessary to explain randomness of activation and an exclusive effect on one chromosome in the presence of two or more X chromosomes. The difficult problem, which has never been adequately addressed, is the spread of the effect from a single activation center to nearly the whole chromosome in the human cell or to a specific part of the X in other species. There appear to be few mistakes in the spreading phenomenon, except when an unusual piece of chromosome is inserted by translocation.

## B. A New Methylation Model

Now that we have evidence that methylation is involved in the inactivation of the X chromosome, as well as many other genetic loci, we should be able to devise a better model. However, let us first look at the evidence that methylation is involved.

In 1980 Jones and Taylor showed that the cytosine analog 5-azacytosine when incorporated into DNA can lead to hypomethylation. The nucleoside, either 5-azacytidine (5-aza C) or 5-azadeoxycytidine (5-aza dC), when present in the medium of cultured cells, is incorporated into DNA and some of the cells which survive have a lower level of DNA methylation. Apparently the analog, which has a nitrogen instead of the usual carbon at the 5-position in the ring can not accept a methyl group, but in addition the analog causes the methylase to be nearly irreversibly bound to the DNA so that many sites remain hemimethylated after one round of replication. If the inhibition of methylation is effective until the DNA replicates a second time the hemimethylated DNA produces one completely unmethylated double-stranded segment and one hemimethylated segment of DNA. The unmethylated segment now has no information at a symmetrical site to allow a maintenance methylase to function, and the methylation pattern is permanently lost at the affected sites.

Mohandas *et al.* (1980, 1981) have produced good evidence that genes on an inactive X can be reactivated by incorporation of 5-azacytidine into cells in culture. Four loci were examined. STS (microsomal steroid sulfatase) is X-linked but is not normally inactivated on the inactive X of either the mouse or human. In the human cell it is located at the distal end of the short arm of the X. Many other mammals have X chromosomes with segments that behave like autosomes in both replication and condensation cycles in addition to the differential segment (inactive part). G6PD (glucose-6-phosphate dehydrogenase), PGK (phosphoglycerate kinase) and HPRT (hypoxanthine phosphoribosyl transferase) are three other genes known to

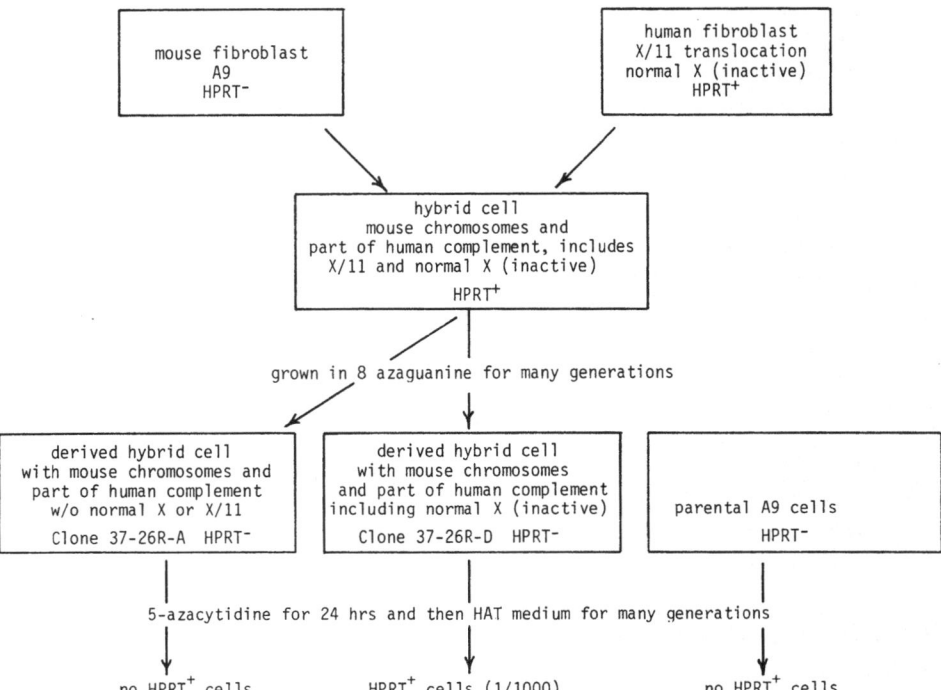

Fig. 13. Mouse fibroblasts (A 9) were fused with human fibroblasts carrying a reciprocal translocation between X and chromosome 11. 11/X and X/11 contain the active X and the normal X is inactive. When the hybrid cells (HPRT$^+$) were grown for many generations in 8-azaguanine which kills HPRT$^+$ cells, clones were derived that were HPRT$^-$. One clone 37-26 R-D has human chromosomes 3, 4, 6, 8, 10, 13, 14, 15, 18, 21, and X (inactive). Clone 37-26 R-A has the same chromosomes except X (inactive) is not retained. When these cells and the parental A 9 cells were treated with 5-azacytidine, the only one to yield HPRT$^+$ cells was the clone 37-26 R-D with the inactive X

be inactivated. MOHANDAS *et al.* used cells of an established mouse line (A 9), deficient in HPRT and fused them to a human fibroblast carrying a balanced X/11 translocation. In the human parental cells, the distal half of the short arm of one X chromosome was translocated to the long arm of chromosome 11. The translocation was reciprocal in that the long arm of 11 was translocated to the X chromosome. The structurally abnormal 11/X was consistently found to be the active X in the human cells and the intact X was the inactive chromosome ($X^{in}$). Among the primary hybrids were cells which had both $X^{in}$ and the derived X (11/X) and were, therefore, positive for all four enzymes. A hybrid clone of this type was treated with 8-azaguanine to select secondary clones in which the 11/X was lost (Fig. 13). Secondary clones with only the inactive X were obtained which did not have the enzymes G6PD, PGK or HPRT. Two of these secondary clones and the original mouse A 9 line were used for 5-azacytidine treatments. One of the hybrids (37-26 R-D) retained human chromosomes 3, 4, 5, 7, 10, 13, 14, 15, 18, 21 and $X^{in}$, while the other, 37-26 R-A, had the same human autosomes but $X^{in}$ was not retained.

Treatment of the 3 cell types with azacytidine yielded clones that are $HPRT^{+}$ from the hybrid with the inactive X but not from the other human-mouse hybrids or from the mouse line, A 9. The frequency of the HAT-resistant colonies (possessing HPRT activity) under optimal conditions (2 μM 5-aza C for 24 hours) was about one per thousand compared with one per million in untreated cells.

Several of the HPRT positive colonies was isolated and tested for the unselected markers. The expression of the human STS gene indicated the presence of the morphologically normal human X ($X^{in}$). The cells maintained the HPRT activity for at least 15 doublings during a 5 week period. Five independent subclones derived from one HPRT, G6PD positive clone maintained both activities for many passages, which indicates that the changes are stable and not dependent on the continuous presence of 5-aza C. Several other similar analogs, 6-azacytidine, arabinosyl-cytosine and 5-bromodeoxyuridine, which do not interfere with methylation, did not yield any HPRT-positive clones.

No tests were made for the state of methylation of any of the X-chromosome genes which were reactivated, but the circumstantial evidence is good that the activation was due to a reduction in methylation of the genes.

Another study of a cloned fragment of X-chromosome DNA failed to produce results indicating that the inactive X chromosome is hyper-methylated. WOLF and MIGEON (1982) studied the methylation of a randomly chosen 28 kilobase segment of X-chromosome DNA by means of cloned probes. Assay was by Southern blot analysis after digestion of chromosomal DNA with Hpa II and Msp I which recognize the same sequence (CCGG) but Msp I is inhibited by methylation of the first C while HpaII is inhibited by methylation of the second C. They did not know whether the segment was homologous with the active or inactive part of the X-chromosome, but since most of the X-chromosome genes are presumably inactivated, it is likely that it is part of the inactive region. The analysis indicated that X-chromosome methylation changes during a sequence of replications in cultured cells, but is not necessarily correlated with the number of X chromosomes or their transcriptional activity. They suggested that in normal human cells the the methylation is less stable, but maintained at a higher level, than when the human X is in a foreign environment of mouse-human hybrid cells. Unfortunately, this study contributes little to our understanding of how methylation is controlled or how it functions. They also failed to confirm the findings of MOHANDAS et al. (1981) that 5-azacytidine incorporation can reactivate the genes G 6 PD and HPRT. One difference which may be significant is that WOLF and MIGEON attempted the activation in normal human skin fibroblasts instead of heteroploid cell lines produced by cellular fusion. Either the number of CpG sites methylated might be less in the hybrids or the normal cells may have evolved ways to restore methylation at some sites even when the maintenance methylation fails to operate effectively. There are also the puzzling cases of de novo methylation of integrated viral genomes (SUTTER et al. 1978, JAHNER et al. 1982), which presumably are produced after integration, but this de novo methylation appears to be limited to cells in the preimplantation embryo. Non-integrated viral DNA which is replicated during infection and viral DNA which is coated as mature viral particles are not detectably methylated. Therefore, one assumes that cells have the capacity to methylate some sequences even when grown in culture. Furthermore, the variations detected by WOLF and MIGEON (1982) were in primary cultures where cellular selection, and perhaps differentiation, is going on (REIS and GOLDSTEIN 1982). One wonders if that could contribute to the variation in stability of the methylation pattern in the 28 kb segment of X-chromosome DNA which WOLF and MIGEON studied.

In a similar study, but one involving a non-sex-linked gene, a result similar to that of MOHANDAS et al. (1981) was obtained by CHRISTY and SCANDOS (1982) and shown to be correlated with a change in methylation pattern. They transformed mouse cells with herpes simplex virus containing the viral thymidine kinase ($TK^+$) gene. The expression of the kinase gene is unstable in the transformed cells. Cell lines were selected that were $TK^-$, although still containing the TK gene, and reversions to the $TK^+$ phenotype were observed at low frequencies in selective medium (HAT) in which only cells expressing TK could survive. The spontaneous rate was about $10^{-6}$. When the cells were treated with azacytidine for 2 days or more, there was a fifteenfold increases of those which expressed the TK phenotype among the surviving cells. Analysis of the methylation patterns of the viral TK gene by the use of the restriction endonucleases HpaII and MspI showed that the TK gene in the cells with viral thymidine kinase was unmethylated. The same loci in the cells without TK activity were methylated, but after azacytidine treatment, the cells which grew in the selective HAT medium were unmethylated.

It is probably safe to predict on he basis of the evidence so far that most methylation patterns will be stably inherited in cells where differentiation and cell selection are not significant phenomena. How then does a change occur and how is the time of the change controlled? This becomes the important question which can explain differentiation. Perhaps it is easier to think of deletion of a pattern of methylation first. How is the loss of a pattern regulated so that a silent gene can become active without its neighbors being affected? The operation of a demethylating enzyme is possible, but its malfunction would very likely be lethal in a cell where much of the regulation is related to methylation patterns. Its action would have to be highly regulated, probably by sequence specificity. It is easier to imagine that site specific inhibitors act at limited sites during certain stages to prevent methylation during one or two replication cycles. Two cycles would be necessary before both chains of a fully methylated DNA would lose the information for restoration of a pattern. Maintenance methylases, which recognize hemimethylated DNA, could restore the fully methylated state at any time before the second replication. However, if a hemi-demethylase removed the methyl group from only one chain, then one replication would be sufficient to completely demethylate one of the two chromatids at a hemi-demethylated site. In the absence of replication such an enzyme would never delete a pattern that could not be restored by a maintenance methylase. It is interesting that the only demethylating activity so far described does not completely remove the labeled methyl groups on the cytosine (GJERSET and MARTIN 1982). It stops at less than 50%, but no information is available to indicate whether the DNA is hemimethylated after the reaction is complete.

# V. Mechanisms of Suppression by DNA Methylation

We have cited a number of studies showing an inverse correlation between the level of methylation and function of a number of genes. What then is the nature of this correlation? The assumption usually is that DNA methylation is the initial event and suppression follows. However, some of the data are just as easily explained on the assumption that function turned off by some other mechanism makes the gene susceptible to methylation and there may not be a cause and effect relationship between methylation and gene activity. As long as the evidence was based only on correlation of the two properties, expression and the unmethylated state, this possibility had to be considered. Recently, however, a cause and effect relationship has been more firmly established.

## A. Expression of the Late Viral Protein of SV 40 (VP 1) Is Reduced by Methylation at One Hpa II Site

As pointed out in Section VI below the 5,243 bp SV 40 genome has eliminated by mutation to TpG most of the CpG sites which it may once have contained. There are only 27 remaining and most of these are clustered near the origin of replication and in the 5′ region of the late transcript (Fig. 14). Only one of these CpG sites is included in a Hpa II site. This site is located near the 5′ end of the late transcript. FRADIN *et al.* (1982) methylated SV 40 DNA with Hpa II methylase. The DNA which remained supercoiled after incubation with the enzyme was then injected into nuclei of *Xenopus* oocytes. Control nuclei were injected with equal amounts (0.25 or 1.25 ng) of unmethylated DNA. Groups of 20 oocytes were incubated in a buffer with $^{35}$S-methionine. After 24 to 72 hours the group of 20 oocytes was disrupted in 1 ml of extraction buffer and aliquots (250 μl) were immunoprecipitated with either anti-SV 40 capsid antiserum, anti-SV 40 T antigen antiserum or non-immune rabbit antiserum as a control. Viral protein (VP-1) increased in direct proportion to the amount of SV 40 DNA injected from 0.02 to 2 ng per oocyte. At 0.25 ng of DNA per oocyte Hpa II methylation caused a marked reduction in VP-1 (Fig. 15) without affecting the production of either of the early proteins, T or t antigen.

This experiment is a striking example of the effect of methylation of a single CpG site in the transcribed DNA located near the 5′ end of the RNA transcript. Unfortunately, with the Hpa II methylase used, it was not possible to methylate the other CpG sites. There are seven more CpG sites clustered downstream of the one Hpa II site and eleven upstream including

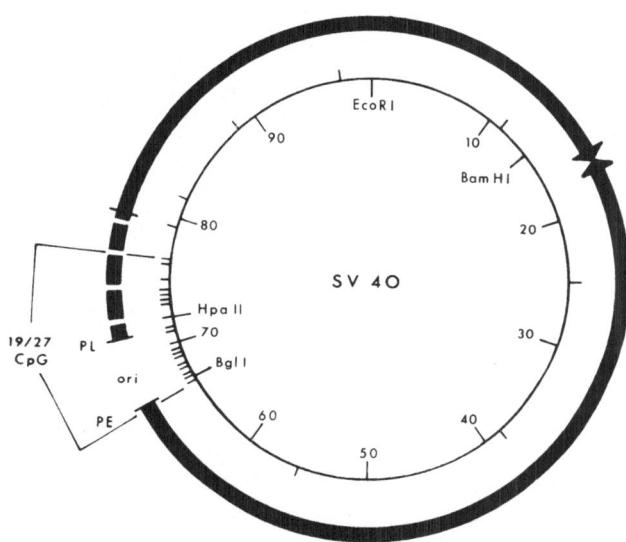

Fig. 14. Map of the SV 40 genome to show the location of the Hpa II site and the CpG sites by the outer projections on the inner ring. The numbers on the inside of the inner ring are the map units measured from the Eco RI site. The heavy outer lines show the early RNA transcript counter clockwise from the Bgl I site and in the other direction the late transcript. (From FRADIN *et al.*, 1982; photograph provided by Dr. JAMES L. MANLEY)

the origin of replication. It will be interesting to know how the methylation of these sites affect the function of the viral DNA. It could be studied by the use of the DNA methylase isolated by SIMON *et al.* (1981 and personal communication) which methylates all of the CpG sites in a supercoiled plasmid or viral DNA. It would also be of considerable interest to see if the integrated SV 40 DNA, which does not produce capsid proteins, is methylated at any of these CpG sites. The only two which can be examined are those in the Hpa II site and the Hha I site just upstream of the Hpa II sequence. Apparently this analysis has not been reported, but it should be quite feasible with available techniques.

Although there is a large cluster of CpG sites just 5′ to the initiation of the early transcript of SV 40, there is no information on their effect, if any, on the function of the T antigen genes. SIMON (personal communication) has

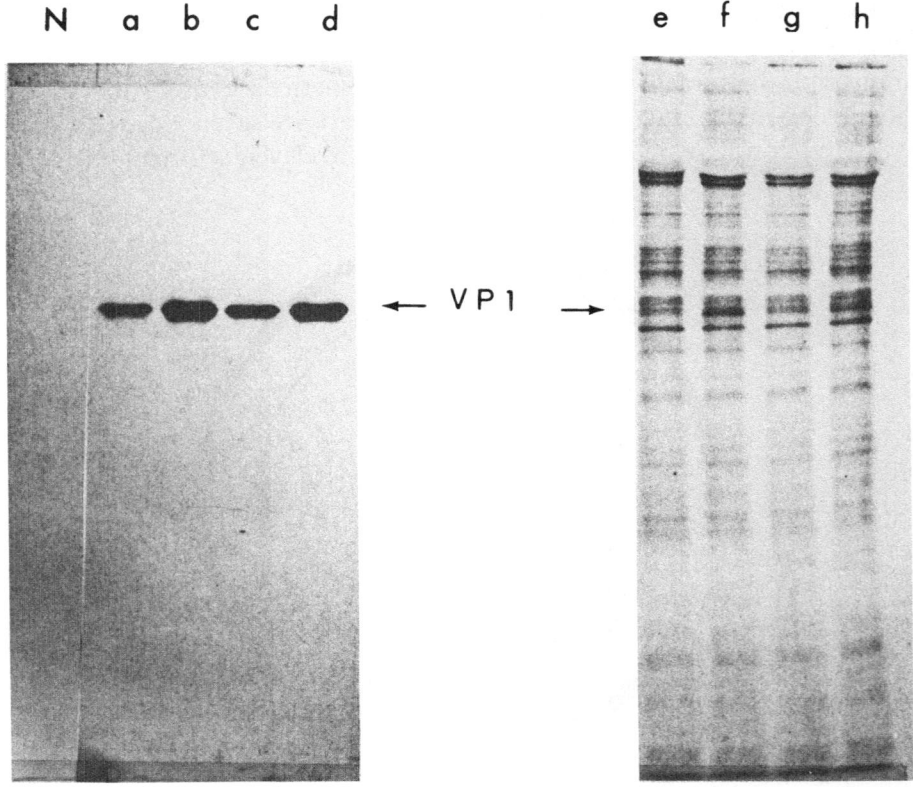

Fig. 15. SV 40 DNA was methylated at the single Hpa II site with Hpa II methylase and microinjected into oocytes of *X. laevis*. Groups of 20 oocytes were injected with either methylated or unmethylated DNA and labeled from 24 to 72 hours post-injection with [$^{35}$S] methionine (2.5 mCi/ml). The proteins were extracted in buffer solution and immunoprecipitated with 10 µl of rabbit serum immunized with SV 40 capsid protein, or 10 µl of non-immune rabbit serum. The capsid protein was released from the immune complexes in the precipitate by heating and one half of the sample was analyzed by polyacrylamide gel electrophoresis. Lane (*a*) shows protein produced by oocytes injected with 0.25 ng of methylated SV 40 DNA, and (*b*) that produced by unmethylated SV 40 DNA. Lanes (*c*) and (*d*) are similar except the oocytes were injected with 5 times as much DNA. Lane N is a control and shows normal rabbit serum immunoprecipitate of an amount of the sample equal to that shown in lane (*d*). Lanes (*e*), (*f*), (*g*), and (*h*) show the separation of proteins in one hundredth of the amount of extract used in lanes (*a*), (*b*), (*c*), and (*d*) respectively, without immunoprecipitation step. Note that the SV 40 capsid protein can be detected in the presence of the other proteins synthesized in the oocytes. (From FRADIN *et al.*, 1982; photograph provided by Dr. JAMES L. MANLEY)

reported that methylation of all of the CpG sites in SV 40 does not block transfection with this DNA. Since the early genes are probably transcribed before any replication of the DNA occurs, the methylated CpGs may not be located where they can affect the expression of the early genes in SV 40. Future investigations on the methylated state of the integrated genome of SV 40 and other papova viruses will be interesting and perhaps revealing of the role of DNA methylation in cells. Unfortunately the analysis of methylation patterns is only possible now at the limited sites which can be probed with restriction enzymes.

The experiment outlined above indicates that there is a cause and effect relationship between methylation and expression of a gene. How then is the effect mediated? Is it an effect on the transcription of the gene? The answer which is emerging is clearly in the affirmative as indicated earlier. The next series of experiments to be described also make that relationship clear.

## B. Transcription of a Cloned Adenovirus Gene Is Inhibited by in vitro Methylation

VARDIMON et al. (1982) made a direct test of the difference in transcription between an adenoviral gene that was methylated at a few of the CpG sites (Hpa II sites) and one that was unmethylated. The E 2a region of adenovirus type 2 encodes a specific DNA-binding protein. In three lines of hamster cells (HE 1, HE 2 and HE 3) transformed by adenovirus type 2, multiple copies of the genome have been demonstrated to persist in the integrated state. Cell lines HE 2 and HE 3 do not have the DNA-binding protein encoded by the viral genome, but cell line HE 1 does. When the methylation pattern of the viral genomes at the Hpa II sites was examined, those in cell lines HE 2 and 3 were methylated and those in line HE 1 were unmethylated. The tests for methylation were made by the use of the Southern blotting technique in conjunction with digestion by Hpa II and Msp I as described previously. In order to make a comparative test the E 2a region (included in the largest Hind III fragment of adenovirus type 2 DNA) was cloned into the plasmid pBR 322 and amplified in E. coli. The plasmid as grown had no methylation at the CpG sites, but some of the DNA was methylated in vitro with Hpa II methylase. These Hpa II sites are shown in a portion of the fragment bracketed by a Hind III site and Eco RI site (Fig. 16 A). The segment from positions 71.4 (Kpn I) to 72.8 (Hind III) was end labeled at the 5′ end and used as a probe for the transcribed RNA from oocytes. After annealing the labeled DNA to the isolated RNA from oocytes injected with the plasmid, the mixture was digested with $S_1$ nuclease. If an RNA transcript had been made off the E 2a gene, the DNA would anneal to it and the [32]P-labeled DNA probe would be protected from digestion. The digest was run by electrophoresis into an acrylamide gel and autoradiographed to reveal the [32]P-labeled probe. DNA fragments 69–71 nucleotides long were protected in lanes c, f and g in Fig. 16 *B*. This shows that RNA transcripts were made on the injected unmethylated plasmid and that the site for initiation was the same as in KB cells productively infected with adenovirus 2. The methylated plasmid (lane d) did not produce any RNA to protect the [32]P-labeled DNA probe. The experiments indicate that methylation somehow prevents transcription of the gene.

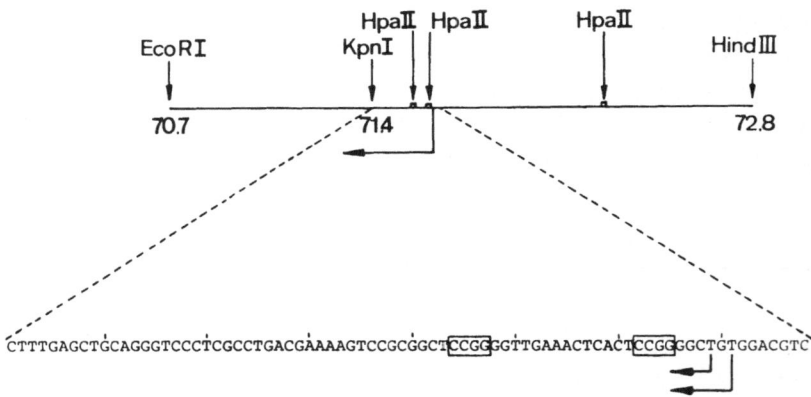

Fig. 16 *A*. Structure of the E 2a region of the adenovirus 2 genome which encodes a DNA-binding protein. The gene may or may not be expressed when the viral genome is integrated in the host chromosome. To test the effect of methylation on transcription of the region, the methylated and unmethylated DNA was injected into Xenopus oocytes and the results analyzed as shown in Fig. 16 *B*. (From VARDIMON *et al.*, 1982; original photograph supplied by Dr. WALTER DOERFLER)

## C. Transcription of a Cloned Human Gamma Globin Gene Is Inhibited by Methylation at the 5′ Region Flanking the Structural Gene

The arrangement of the human beta-related globin genes have been described earlier (Section III). There are five active genes arranged in the order of their expression in development and two pseudogenes. Of the five which are transcribed, the transcription runs from the epsilon gene, which is expressed in early embryos, to the two gamma genes, Gγ and Aγ, which are coordinately expressed in the fetus, and finally to the two adult globins, the delta chain and the beta chain. These have been cloned as a series of overlapping cosmids in a stretch of DNA spanning about 120 kb. Introduction of a globin gene into mouse L-cells in vectors, which also carry the *Herpes simplex* virus thymidine kinase (HSV tk) gene to facilitate selection of transformants, results in a constitutive expression of the introduced globin gene. The recipient cells are TK⁻ and if competent for

a b c d e f g h i

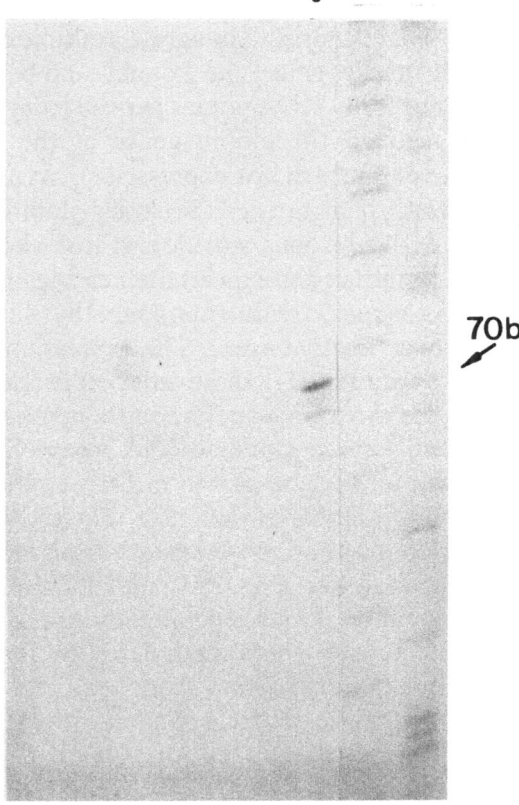

70b

Fig. 16 *B*. Autoradiograph showing effect of DNA methylation on the transcription of the adenoviral gene from the E 2a region. The Hind III A fragment of the viral genome with and without methylation by Hpa II methylase ($C^mCGG$) was injected into oocytes of *Xenopus laevis* where the gene can be transcribed. After incubation for 24 hours the RNA was extracted from the oocytes and the mixture hybridized with a $^{32}$P-labeled probe consisting of the single strand complementary to the RNA transcript and extending over the region from the Kpn I site to the Hind III site of the E 2a region (Fig. 15 *A*). After hybridization the mixture was digested with nuclease $S_1$ which destroys single stranded RNA or DNA. The part of the RNA transcript that overlaps with and pairs with the probe, consisting of 69–71 nucleotides, is protected and shows up when the digest is separated by electrophoresis. Appropriate controls were included and an autoradiograph of the gel is shown. Lanes: *a* nuclear RNA from K B cells isolated 12 hours after infection with Ad 2 (the hybrids were not treated with nuclease $S_1$); *b* no RNA; *c* total RNA extracted from *Xenopus* that had been microinjected with unmethylated cloned Hind III A fragment of Ad2; *d* total RNA from oocytes that had been microinjected with Hpa II methylase-treated cloned Hind III A fragment of Ad 2 DNA; *e* RNA from oocytes that had not been microinjected; *f* nuclear RNA isolated 12 hours after infection of KB cells with Ad 2; *g* cytoplasmic RNA isolated 12 hours after infection of KB cells with Ad 2; *h* and *i* size markers from a routine sequence analysis experiment. (From VARDIMON *et al.*, 1982; original photograph supplied by Dr. WALTER DOERFLER)

transformation, they will take up and integrate multiple copies of the globin gene as well as the TK marker gene. The gamma- and beta-globin genes are expressed at levels up to 5,000 RNA copies per cell from a few copies of the cosmid DNA, even though the globin genes of the transformed cell's endogenous, natural genome are not expressed (FLAVELL *et al.* 1982).

FLAVELL *et al.* (1982) also introduced the cloned globin genes into another TK cell line, this one derived from a murine erythroleukemic cell line (MEL cells). These cells can be induced to express their endogenous α- and β-globin genes by a variety of chemical treatments. Dimethyl sulfoxide (DMSO) is perhaps the best known effective agent. The expression of the transfected human globin gene was examined in these cells before and after induction of endogenous globin gene expression. Before induction a low level of human beta-globin mRNA and gamma-globin mRNA was produced in most of the derived cell lines. The human beta-globin mRNA could be induced along with the endogenous mouse beta-globin gene. Therefore, the human beta-globin genes appear to respond to the same transacting signals as the endogenous mouse beta-globin gene, but neither the introduced gamma-globin gene nor the gamma-globin gene already in the cells responded.

They then tried introducing genes methylated by the method of STEIN *et al.* (1982) *i.e.* hemimethylated DNA from segments cloned into M 13 (single strand phage) and made double-stranded with DNA polymerase I in which 5-methyl-dCTP is was substituted for dCTP in the reaction mixture.

---

Fig. 17. The pattern of methylation inherited in the human $A_\gamma$-globin genes used to transform four clones of mouse-L cells. The map of the gene and the position of the Hind III, Hha I and Hpa II sites are shown below the autoradiograph. The primer used to partially methylate the gene is indicated under the map. The globin gene had been cloned into the double stranded, replicative form of the single-stranded phage, M13. The single-stranded form of the phage with the globin gene was then used as a template for the synthesis of a hemimethylated DNA. By the use of an appropriate primer segment, DNA polymase I and the precursors, $d^mCTP$, dATP, dTTP and dGTP, different parts of the gene could be hemimethylated. The autoradiograph shows the methylation pattern inherited by globin genes in several clones of L cells. The first 3 lanes shows the pattern of digestion of the unmethylated M 13 vector, containing the $A_\gamma$-globin gene ($M_\gamma 1$). Clone 9 was transformed with unmethylated DNA and shows the same pattern of digestion as the vector borne gene, $M_\gamma I$. Clone 10 was transformed with DNA methylated at all of the cytosines and has retained the methylcytosine at all Hha I and Hpa II sites. Clones 11 and 13 were transformed with partially methylated DNA (no methylcytosine in the primer region). Clone 11 has lost its methylation since its digestion pattern is similar to unmethylated Clone 9, but Clone 13 has retained the methylation at Hha sites 1 and most of the Hpa II sites 1. Clone 9 produces γ-globin RNA transcripts, but Clone 10 does not (data not shown). Clone 11 produces transcripts but Clone 13 does not. (From FLAVELL *et al.*, 1982; copyright by Cold Spring Harbor Laboratory, 1982; original photograph provided by Dr. M. BUSSLINGER)

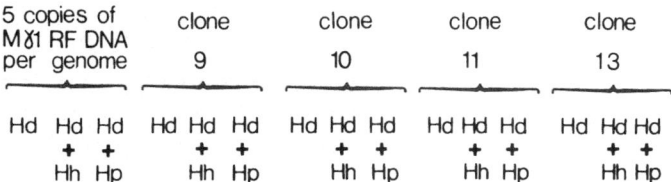

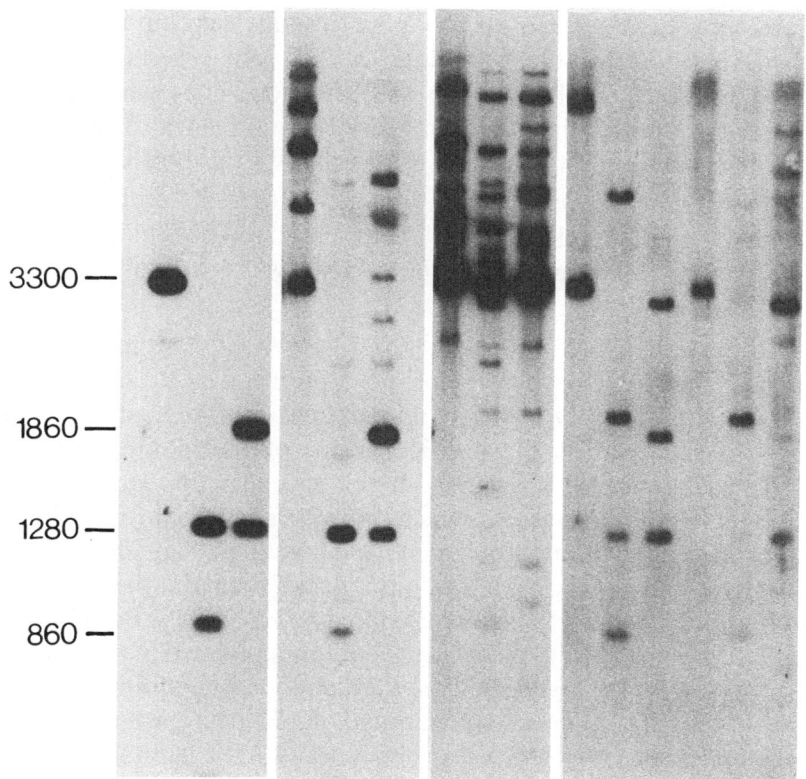

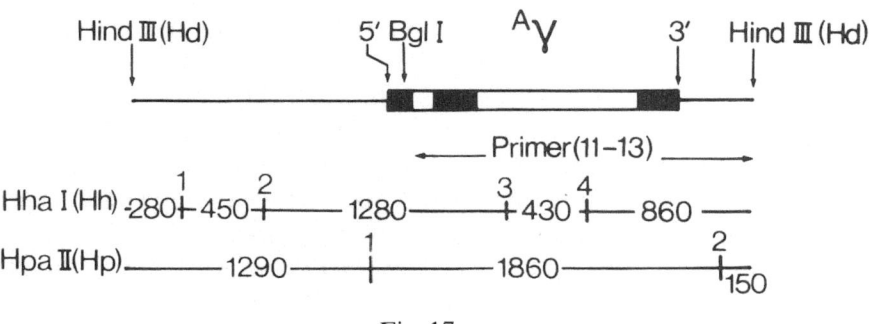

Fig. 17

When this DNA was used for transfection, the integrated globin genes maintained the methylated CpG sites, but lost the other methylated cytosines because they are not maintained during replication. When TK$^-$ L-cells were transformed with the viral vector containing methylated alpha-globin genes and unmethylated beta-globin genes, the cells most likely to have the two globin genes could be selected in HAT medium by co-transforming with a vector containing the herpes TK gene. Those cells which are competent to take up DNA will usually take up and integrate whatever DNA is available, including the TK gene. Tests then can be performed by Southern blotting to identify those lines which have the methylated and unmethylated globin genes. When this was done, only the unmethylated beta-globin gene was transcribed in the L cells.

Similar experiments were carried out with either the alpha or gamma globin gene unmethylated along with the methylated beta globin gene as a control. In each case the unmethylated gene was transcribed while the methylated gene was not.

Since the methylation by synthesis of a second strand with DNA polymerase I requires a primer, the primer region can be left unmethylated. This allowed FLAVELL et al. (1982 and personal communication) to study the effect of methylation of specific regions of the globin genes. They prepared a partially methylated gamma globin gene in this way. The primer was the Bgl I/Hind III fragment which anneals from the position + 100 in the first exon of the gamma-globin gene to a position 300 nucleotides to the 3' side of the A-gamma-globin (A$\gamma$-globin) gene. When this primer was used the gamma-globin gene was hemimethylated in the entire 5' flanking region and the first 100 nucleotides of the gene, but unmethylated in the main body of the gene (Fig. 17). This partially methylated gamma-globin gene, along with the herpes TK gene and an unmethylated beta-globin gene as an internal control, was used to co-transform mouse L-cells. In two clones the partially methylated gamma-globin genes were not transcribed, but in a third clone isolated in the same experiment, the gamma-globin gene was transcribed (Fig. 17). The genes were then analyzed for the pattern of methylation. The non-transcribed genes had both maintained the methylation at the two HhaI sites and the HpaII site just 5' to the Bgl I site (Fig. 17). On the other hand, the transcribed gene had lost the methylation at these sites and presumably the other CpGs in that portion of the gene. This may have occurred by repair synthesis during integration, since such a loss is consistent with all other transformation experiments in which maintenance of a methylation pattern is not perfect. However, this clone serves as a good control. The transcription is blocked by methylation of the 5' flanking region and the first part of the coding region. Which of these CpG sites are effective is, of course, not yet known. However, other transformants have been obtained in which the methylation was complementary to that described above. The main body of the gene was methylated and the 5′ flanking region down to the + 100 position was unmethylated. In these transformants the gamma-globin gene was transcribed as well as in the unmethylated controls.

In one other type, the M 13 vector portion of the DNA was hemimethylated but the entire Hind III segment remained unmethylated. Since this gene was transcribed as well as the clone transformed by unmethylated DNA, the methyl groups in the adjacent M 13 DNA sequences do not exert any long distance effect; the mCpGs in the Hind III segment are sufficient to exert the full effect.

## D. The Mechanism by Which Methylated CpG Sites Inhibit Transcription

Although the evidence is now in favor of an inhibition of gene expression at the level of transcription, the mechanism by which the methylation acts is still obscure. In all of the cases cited above the methylated or unmethylated DNA is operating in a cellular environment where all of the proteins of the nucleus are available to interact with the DNA. Experiments with isolated DNA transcribed in vitro do not appear to be affected by the CpG methylation of Hpa II sites in the hemoglobin genes (SHEN *et al.*, personal communication). Therefore, it seems likely that the effect is mediated through the binding of molecules other than RNA polymerase. There is certainly no firm answer that can be given at this writing, but since the inhibition operates in the nuclear environment and on chromatin, there are many possible proteins that could be involved. The point has also been made by many investigators that although methylation can prevent gene expression, the deletion of a methylation pattern is not in itself sufficient for gene expression or even for transcription. It appears more likely that cells have utilized methylation as a mechanism to selectively prevent the transcription of genes in a nuclear environment where all other factors for transcription may be available, but transcripts from certain sites would be disruptive to the operation of the cell. In some cases it might just be a wasteful operation in terms of energy conservation, but in others it might be that expression would be very disruptive for the metabolism or life of the cell. Certainly the suppression of viruses can be viewed as a very beneficial bonus, even if that was not a primary factor in the evolution of methylation mechanisms.

One problem that we always come back to, in considering the usefulness of the methylation pattern in controlling which genes are free to function, is the failure to find any such patterns in the insects, particularly *Drosophila* in which the search has been extensive. Of course, it is always possible that some other modification than those at CpG sites could be involved in insects and have thus far eluded detection. However, this seems less likely than that some other mechanism for turning off parts of the genome may be available to arthropods and perhaps some other groups of animals.

## E. How Are Methylation Patterns Established and Maintained?

The establishment of a rather uniform pattern of methylation of all CpG sites could occur in the germline by some type of de novo methylase which operates rather nonspecifically with respect to flanking sequences. The methylases in most somatic tissues could then be maintenance type methylases which replicate a pattern based on the symmetrical methylcytosines on the parental chains after each round of replication and after repair synthesis which replaces a 5-methylcytosine with cytosine. Depending on the efficiency and fidelity of the maintenance enzymes such a system might work reasonably well. Unmethylated sites could arise by methylase inhibitors which would require considerable sequence specificty to uncover and release the correct genes for expression in differentiating cell lines. The high level of methylation of DNA in spermatozoa seems to fit this model reasonably well if we assume that DNA from the egg is also highly methylated.

We have already discussed several studies which support such a model, but there are clearly some important exceptions which do not fit the model. First of all, cells must grow and function in the germline and those genes which are transcribed could not be maintained with a high level of methylation, if it regularly inhibits transcription. There is, of course, an alternative. If non-specific methylases operated in the late stages of gametogenesis after all necessary transcripts were made the total genome could be methylated at CpG sites to start the next generation. For some classes of genes and in some species, the de novo methylation could be delayed until the zygote or very early embryo. Differentiation would then be a process of selective inhibition of maintenance type methylases at various stages. Let us review the data on methylases and maintenance of methylation patterns and then reconsider a model for establishment of specific patterns and their maintenance through development and differentiation.

## F. The Specificity of Methylases and the Maintenance of Methylation Patterns

### 1. Prokaryotic Type II Methylases

Prokaryotic DNA methylases are of three general classes. The best characterized are modification enzymes corresponding to the Type II restriction endonucleases. These enzymes are all de novo methylases with a high sequence specifity. Both chains are methylated at symmetrical position on either adenines or cytosines, depending on the specificity of the methylase and its recognition sequence. The molecules are dimers which rapidly and completely methylate both chains at the recognition sites. They have little if any preference for hemimethylated DNA, and therefore provide few clues for the maintenance of methylation patterns in eukaryotic DNA. The corresponding endonucleases do not cleave the hemimethylated DNA, but some of them, for example Hae III (GRUENBAUM et al. 1981), can nick the unmethylated chain. The enzymes require only S-adenosylmethionine in vitro and can be used in the presence of chelating agents, such as EDTA, which inhibit most nucleases. Their principal use relevant to our discussion has been to methylate eukaryotic DNAs which can then be introduced into cells to see how well the pattern is maintained. The most useful one has been the Hpa II methylase which was initially isolated and characterized by MANN and SMITH (1977). The CCGG site is methylated on the inner cytosine. A related methylase from *Morexella* recognizes the same site but methylates both cytosines (JENTSCH et al. 1981). A methylase with the same specificity has also been isolated from *Bacillus subtilis* infected with phage SP. A similar but more limited reaction by a cell-free extract of *B. subtilis* (strain Q) methylates only the outer cytosine which makes the DNA resistant to Msp I. These methylases are useful for methylating eukaryotic DNAs for in vitro studies or for introduction into cells, but give few if any clues concerning the mechanism of "maintenance" enzymes in eukaryotes. On the other hand, the Type I and Type III modification enzymes may have some characteristics of interest in this respect.

### 2. Type I and Type III Methylases

Type I methylases as mentioned earlier modify sequences such as 5'-TCA(N8)TGT (*E. coli* B) and 5'-AAC(N6)GTGC (*E. coli* K). Hemimethylated DNA is rapidly methylated by these enzymes and is not susceptible to either single- or double-stranded scission by the corresponding endonucleases. The enzymes require only S-adenosylmethionine (SAM), but are stimulated by ATP and $Mg^{2+}$ (YUAN 1981). The unmodified DNA is methylated by the enzyme in the absence of ATP but only very slowly after a lengthy incubation in vitro. The enzyme is a multiunit complex (400,000 M.W.) which has three enzymatic activities, a restriction endonucleases that cleaves unmodified DNA in the presence of ATP, SAM and $Mg^{2+}$, a modification methylase and an ATP hydrolyase that is coupled to restriction in a non-stoichiometric way in vitro.

The Type II methylases are similar to the Type I enzymes in that a large complex can modify and restrict DNA, but there are some important differences. Whereas Type I requires SAM for endonuclease activity, Type III enzymes are only stimulated by that cofactor. Type II endonucleases produce distinct fragments while Type I produces a heterogeneous population of fragments at unpredictable distances from the recognition sites. The enzymes modify and cleave DNA simultaneously, but at different sites; if modified the site is not cleaved.

The Type I modification enzymes have some properties one might expect for maintenance enzymes in eukaryotes, but a methylase coupled with endonuclease activity has not been demonstrated in a eukaryotic cell. A methylase of this type would be expected to modify sites very slowly when unmethylated, but after the initial methylation, the newly replicated DNA (hemimethylated) would be rapidly methylated on the new chain. The major difference is that these bacterial enzymes have specific recognition sites involving flanking sequences around the methylated base, while the eukaryotic counterparts would be expected to have very limited recognition specificity; CpGs with nearly any flanking sequence seems to be a substrate for the de novo methylases of eukaryotes. The recognition sites if more extensive than the unmethylated CpGs, are probably dispersed repetitive sequences located some distance from the target CpGs. The enzymes might recognize a sequence common to a number of genes under coordinate control and "walk" along the DNA methylating all CpGs until bumped off by another signal sequence. As mentioned earlier the Alu-like sequences are good candidates for such a function since they appear to be origins for replication. Changes in methylation patterns appear to be restricted to cells that undergo division or at least DNA replication, which is usually necessary before division.

### G. Properties of Eukaryotic Methylases

Direct evidence for a methylase that has a great preference for hemimethylated DNA was reported by GRUENBAUM et al. (1982). A partially purified preparation from mouse ascites tumor cells methylated hemimethylated DNA hundredfold faster than it did the unmethylated control DNA. The substrate was hemimethylated DNA that had been prepared on single-stranded $\emptyset$X 174 DNA by primed synthesis with E. coli DNA polymerase I. Hae III endonuclease was used to nick the unmethylated chain and that chain only was labeled by nick translation with one of the four alpha-$^{32}$P-labeled deoxynucleoside triphosphates. By using all four nucleoside triphosphates separately and preparing four different substrates with the unmethylated chain labeled, GRUENBAUM et al. were able to carry out a nearest neighbor analysis and show that the methylase activity put methyl groups on cytosine only in CpG sites. Although other reports have indicated some methylation of the first cytosine in CCGG sites (VAN DER PLOEG and FLAVELL 1980, VAN DER PLOEG et al. 1980, and ROY and WEISSBACH 1975), these sites were not detectably methylated by the enzyme activity from the ascites tumor cells.

The specificity of methyl transferases may be slightly different in vivo than when purified and used in vitro. For example, SIMON *et al.* (1982) and SIMON (personal communication) purified a rat liver methylase which methylated double-stranded DNA, single-stranded DNA, and hemimethylated DNA about equally well. The synthetic polymers (dC-dG)n and (dC-dI)n were completely methylated by the preparation of highest purity, about 6,000-fold purified. Three other polymers, (dC-dA)n, (dC-dT)n and (dC-dC)n, were not methylated, but some CpA, CpT and CpC sites were methylated in natural DNAs in addition to the usual CpGs.

The purified enzyme does not "walk along" double-stranded SV 40 DNA because there was non-coordinate methylation of the two Hha I sites as well as the adjacent Hha I and Hpa II sites; the latter two are very close together. The enzyme coordinately methylates the two opposite, or symmetrical CpGs, on both strands at the Hpa II site in SV 40, perhaps because the enzyme is a dimer. This result agrees substantially with earlier reports, one with a 270-fold purified preparation from HeLa cell nuclei (ROY and WEISSBACH 1975), one with a preparation from Krebs II ascites cells (TURNBULL and ADAMS 1976), and a third with a preparation from rat hepatoma cells (SNEIDER *et al.* 1975). In one earlier study DRAHOVSKY and MORRIS (1971 a, b) obtained evidence that a methylase from mammalian cells may "walk" along the molecule from some preferred binding sites. They found a strong DNA-enzyme complex, which once formed, was resistant to dissociation by salt or competing DNAs.

Much remains to be learned about methylases. The more highly purified preparations may have altered specificity with respect to both recognition sites and processive action, *i.e.,* "walking" along the DNA. The enzymes isolated so far usually operate better, or at least as well, on single-stranded DNA than on double-stranded substrates. However, if the molecules are dimers which coordinately methylate the two chains at a given site, the natural substrate is likely to be double-stranded DNA. Only a few of the preparations have a preference for a hemimethylated substrate, which would be expected to be the usual substrate in DNA that has recently been replicated. Other subunits which affect these properties of the enzyme complex could be lost in purification. The only methylase which exhibited a marked preference for hemimethylated DNA (GRUENBAUM *et al.* 1982) was a partially purified preparation from mouse ascites cells.

## H. The Maintenance of Methylation Patterns Imposed in vitro

The maintenance of a pattern of methylation can in principal be studied by one of two methods. DNA which has been methylated with a restriction type methylase, such as Hpa II or Eco RI methylase, can be introduced into a cell by microinjection or by transformation/transfection and the persistence or loss of the pattern can be determined after several or many cell divisions. Alternatively, the methylated base can be introduced by synthesis of a new strand on single-stranded DNA and replacement of all of the cytosines with 5-methylcytosine. The DNA is then introduced into a cell by transformation and after many cell divisions the DNA is examined to see which 5-methylcytosines have been retained.

The first technique has been used by HARLAND (1982) and by APPLEGATE and TAYLOR (unpublished). HARLAND methylated bacterial plasmid DNA with Hpa II methylase and injected these plasmids into unfertilized eggs of *Xenopus laevis*. After the DNA had replicated through one or two rounds, as indicated by incorporation of the density label, bromodeoxyuridine, the DNA was isolated and cut with Msp I and Hpa II endonuclease to see how well it had retained the pattern imposed by the Hpa II methylase. Msp I cleaves the DNA at all Hpa II sites before and after methylation as well as after injection. However, methylation prevented cleavage of the DNA by Hpa II after it had replicated two times; therefore HARLAND assumed that the pattern had been maintained. APPLEGATE and TAYLOR (unpublished) methylated the plasmid DNA in a similar way, but used two methylases, Hpa II for certain CpG sites and Eco RI which methylates certain adenines. The plasmid DNA contained a replication origin from *Xenopus laevis* genomic DNA (WATANABE and TAYLOR 1980). The semiconservative dilution of the methyl-adenine allowed part of the plasmids to be digested by Eco RI endonuclease, but the Hpa II methylation was maintained in all of the plasmids. When SV 40 DNA was methylated in the same way and injected into the *Xenopus* eggs, it replicated and maintained the Hpa II methylation. However, when used to transfect monkey kidney cells, the natural host, the methylation pattern of the single Hpa II site was diluted at the same time as that at the Eco RI into. Even though the SV 40 DNA is replicating in nuclei of monkey kidney cells where methylases are active on other DNA, the viral DNA somehow escapes methylation; the methylation of CpG sites are not maintained during infection.

HARLAND (1982) also injected DNA that was methylated on one chain only, *i.e.*, hemimethylated DNA. In the *Xenopus* egg this DNA had the methylation pattern copied on the unmethylated strand in the absence of replication, but less efficiently than during replication. These experiments show that hemimethylated sites contain the information for transmission of the pattern to progeny molecules and that the maintenance does not require either integration into the chromosome or an origin of replication homologous to those of the cell in which the plasmid is replicating. However, these experiments can provide only limited information on the fidelity of the maintenance methylases. The fidelity of the process requires a methylated gene that can be assayed at mutant frequencies after many rounds of replication. Several experiments are now available that give some information on this topic. POLLOCK *et al.* (1980) methylated the HpaII sites in the plasmid, pBR 322, which contains a cloned fragment of herpes virus, including the complete thymidine kinase gene. It was then used to transform a L-TK$^-$, APRT$^-$ clone of mouse cells and the TK$^+$ cells were selected and assayed 25–50 cell generations later for the methylated CCGG sequences. Only about 10% of all transformed clones contained the methylated sequences, but modification was inherited in a stable condition once those clones were established. On the other hand, cells transformed with unmethylated plasmids acquired the TK gene, but despite the fact that recipient cells had 70% of their genomic CCGG sites methylated, no methylated plasmids or TK sequences were detectable. WIGLER *et al.* (1981) used similar methods to study the retention of the HpaII methylated sites in pBR 322 containing the complete chicken TK gene. After 25 cell generations the state of the DNA was assayed with methylation sensitive endonucleases, blot hybridization and autoradiography. The data indicate that the methylation pattern was inherited, but not with 100% fidelity. The loss of methyl groups probably occurred early, before or during the integration, for they reported evidence that methylation reduced the efficiency of transformation. However, the methylation was not eliminated by an all or nothing mechanism, for many sequences introduced had some HpaII sites accessible to HpaII endonuclease and other sequences were resistant.

Another method of preparing methylated DNA has been more successful in getting transfecting DNA integrated in the methylated state (STEIN *et al.* 1982). STEIN *et al.* prepared methylated DNA from the replicative form of ⌀X 174, pBR 322 and pGH by use of HpaII methylase. They also methylated ⌀X 174 by primed repair synthesis on the single-stranded viral DNA (+ strand) with 5-methyl deoxycytidine triphosphate (5mdCTP) substituted for dCTP in the in vitro reaction. In the resulting replicative form, all cytosines in the (—) strand were substituted with 5-methylcytosine, except the five cytosines in the primer. This DNA was completely resistant to all restriction endonucleases that are inhibited by methylated cytosine. Among these are HpaII (CCGG), MspI (CCGG), HaeIII (GGCC), AluI (AGCT), HhaI (GCGC), EcoRII (CCA/TGG), and SocII (CCGCGG).

When the HpaII methylated DNA was used for transfection of Ltk$^-$, aprt$^-$ mouse cells, along with the cloned thymidine kinase gene in pBR 322, to identify the transformed clones, the maintenance of methylation on the ⌀X 174 DNA was less than 100%. Among 10 clones isolated after co-transfection, all were either totally or partially methylated at HpaII sites. In contrast 5 clones isolated after transfection with unmethylated DNA showed no detectable methylation at any CCGG sites in the non-selected marker, ⌀X 174. The transfecting ⌀X 174 DNA had more than 98% of its HpaII sites methylated, but the DNA isolated about 25 cell generations after transfection had only partial maintenance of the pattern of modification. The digestion patterns with MspI and HpaII indicated that dilution of methylated cytosine had occurred at random in the multiple copies of $\emptyset$X174 integrated into the cellular genomic DNA.

By contrast the $\emptyset$X174 DNA, with one strand completely substituted with 5mC except the primer region, showed more efficient retention or maintenance of the mCpG sites; all other methyl cytosines were diluted and not maintained as the DNA replicated. All of the HpaII sites in integrated $\emptyset$X174 were very resistant to the corresponding endonuclease, but susceptible to MspI. This shows that the first methylcytosine in the CCGG sites was not maintained because MspI is inhibited by that type of methylation; however, in the same sites the second 5mC was maintained. The sites for HhaI and SacII also remained resistant to these endonucleases; each contains mCpG which confers resistance. On the other hand the DNA was cleaved by AluI, HaeIII and EcoRII, all of which are inhibited by 5mC in positions that are not part of a mCpG doublet. These experiments provide strong evidence for a maintenance type methylase which specifically maintains mCpG sites, but no other 5mC in detectable amounts. Control experiments were also included which would have detected any significant de novo methylation of the integrated viral DNA; none was detected.

### I. Deletion of Methylation Patterns During Differentiation

If most functional genes are methylated in the sperm and egg or become methylated in the early embryo, the first step in the turning on of genes during differentiation is the deletion of at least that part of the methylation which prevents transcription. Recently, one report of a demethylating activity has been published (GJERSET and MARTIN 1982). Since the activity was detected only in a crude cellular extract, its properties are difficult to predict, but if it operates to produce the variety of pattern changes which arise during differentiation its specificity with respect to flanking sequences are difficult to imagine. As mentioned previously, if it produced only hemimethylated DNA it would not destroy patterns at random and could allow the deletion in only one of the daughter cells following a replication of the hemimethylated DNA. Of course, repair methylation and/or action at specific sites would still be required to produce the degree of specificity observed during development. Without such an enzyme semiconservative replication would require two rounds, and presumably two cell cycles, to achieve the unmethylated state. Yet, we observe instances where a cell divides and produces one differentiated cell and one stem cell capable of repeating the process again and again. However, no specific information is available concerning the state of methylation in the two cells mentioned above. BEACH and PALMITER (1981) did find that the repressed metallothionen-I gene in mouse cells, began to function before two cell cycles could have occcurred when cells were treated with azacytidine. The explanation is not known, but gene amplification or one replication of a hemimethylated gene could explain the observation.

We will assume that de novo methylases operate only at limited times in development as indicated by the de novo methylation of integrated retroviral genomes discussed previously (JAHNER et al. 1982). Differentiation or determination after implantation of the embryo in mammals then becomes a matter of selective inhibition of the maintenance methylases. We have few, if any, clues concerning such inhibitors. They could be effective only if sequence specificity, perhaps of flanking regions of genes, is very precise. To reduce the number of molecules required we may assume that the recognition sites are dispersed repeats that flank families of genes that are required for some steps in differentiation.

Two potential inhibitors are 5-fluorodeoxycytidine (5-FdC) and azacytidine or azadeoxycytidine. Most cells exclude or degrade such intermediates rather effectively. Thymidine analogs can be taken up by most cells and they can substitute for a high proportion of the thymine residues in DNA during synthesis, but it is much more difficult to get cytosine replaced with analogs that have a substitution at the 5-position (CHAMBERS and TAYLOR 1982). However, azacytidine is phosphorylated and reduced to azadeoxycytidine diphosphate, which after conversion to the triphosphate can replace deoxycytidylate in DNA to the extent of 3–5% of the cytosine residues in the newly synthesized strand. Under these conditions the methylation of the new chain is almost completely blocked (JONES and TAYLOR 1981). The reason for the complete block is not only that the azadeoxycytosine residues can not accept a methyl group at the 5 position, but the methylase is severely inhibited. There is more demethylation than can be accounted for by the incorporated analog.

The other analog, 5-fluorodeoxycytidine, which is also incorporated into DNA to about the same extent as azadeoxycytidine, is less effective in blocking methylation as measured by its effect on cellular differentiation (JONES and TAYLOR 1980). The addition of a methyl group at the 5 position would almost certainly be prevented by the fluorine, but the less effective binding of the enzyme and consequent blockage of the methylation at other sites by the presence of azacytidine is probably the explanation for the different effects of the two analogs. The incorporation of either analog into the DNA chain will result in the deletion of the methylation pattern at the CpG site involved. Whether it will affect other sites depends on the mode of operation of the methylase. If the methylase binds flanking sequences, origins of replication for example, and "walks" along the DNA until bumped off, a small amount of analog that either irreversibly binds the enzyme or "bumps it off" prematurely could have major effects on maintenance of methylation. If the methylase operates by random attachments, the effect would be enhanced primarily by abnormal binding.

The data are insufficient to give clues concerning the mode of action of the methylases, but there would appear to be a need for both kinds of methylase, the de novo and the maintenance type. During DNA repair the new DNA chains are remethylated (DRAHOVSKY et al. 1976; HILLIAR and SNEIDER 1975, KASTEN et al. 1982) and these enzymes might operate by random attachment at hemimethylated sites. Methylation immediately following replication from an origin might be more efficient if the methylase attached at the origin and moved along behind the fork. The Okazaki fragments are reported to be unmethylated, but the methylase appears to operate quickly after replication for most of the sites in DNA (ADAMS 1974, DRAHOVSKY and WACKER 1975).

A recent kinetic study indicates that all of the hemimethylated sites on the newly replicated chain become fully methylated within 2.5 minutes. The new chains begin to be methylated within one minute of replication (GRUENBAUM et al. 1983) and the half time for completing the methylation is about 2.5 minutes in mouse L-cells. Since the replication fork moves about 1,500 to 3,000 bp per minute, the methylating enzymes may follow the replication fork rather closely. An earlier study (DRAHOVSKY and WACKER 1975) indicated that the Okazaki fragments behind the fork are not yet methylated. This may mean that the maintenance methylase operates along with or just following the enzymes that remove the primers of Okazaki fragments and ligate the fragments. Another study indicates that the steady state of methylation during replication is reached in less than two hours, the first time point measured, but that repair replacement of 5-methylcytosine is considerably slower. Replacement requires more than 20 hours in log phase cells damaged by ultraviolet light, but the replacement is only a little over one half complete after 3 days in confluent non-dividing cells (KASTEN et al. 1982).

HARLAND (1982) reported that when hemimethylated DNA was injected into *Xenopus* eggs and analyzed before replication had occurred, the unmethylated chain had become partially methylated at scattered sites. This is indicative of a repair type enzyme activity which may operate in addition to a methylase to insure complete retention of a methylation pattern following replication.

The most likely mechanism for the deletion of a methylation pattern would be the competitive binding of a protein to a site recognized to be an origin by a methylase that scans the newly replicated DNA. If the methylase were inhibited in this way, the pattern of deletion of methylation would be expected to involve whole replication units rather than individual CpG sites. This mode of deletion of methylation appears to fit some of the changes revealed by HpaII-MspI digestion of genes and their flanking regions, but the data would have to be much more complete to support the mechanism mentioned above or to suggest a different one. The deletion of methylation patterns by the use of an inhibitor such as azacytidine would be expected to be more random. The possible role of recognition sites involving the dispersed repeats such as Alu sequences will be considered in more detail later. If the deletion of a methylation pattern occurs by the attachment of a protein to an origin of replication, we would expect that origins have common sequence but vary enough to allow functionally related groups of genes scattered through the genome to be activated by one type of inhibitor molecule. The search for a mechanism for deletion of methylation patterns in one or two genes of a cluster such as the β-globin cluster is certainly an attractive part of any study of the correlation between methylation and differentiation.

Perhaps just as important is a search for mechanisms to methylate other families of genes de novo. This would be in contrast to a non-specific methylation of all CpG sites at some stage in development. There is no evidence that such a complete methylation ever occurs. Therefore, the de novo methylases are likely to recognize certain widely dispersed flanking sequences and scan the DNA to another sequence that would terminate the scan. The methylation of proviral sequence in the early embryo (JAHNER 1982) is a possible example of such a mechanism.

# VI. Evolution, Stability and Regulation of Methylation Patterns

## A. The Evolutionary Aspects of DNA Methylation

### 1. 5-Methylcytosine Residues Are Hotspots for Mutation in Bacteria

BENZER (1961) demonstrated large variations in the mutability of certain sites in phage T 4; he called the highly mutable sites "hotspots". COULONDRE et al. (1978) studied the effects of 5-methylcytosine on mutation rates in the lacI gene of E. coli, and showed that several hotspots for base substitutions involve the transition from a G:C base pair to A:T in the lacI gene. The hotspot occurred at the second cytosine in the sequence CCAGG which is known to be methylated at this position by the product of the mec gene in the strain of E. coli K 12 used in the study. To test the hypothesis that the hotspots were the result of the methylation rather than some other characteristic of the sites a new CCAGG sequence was selected. From an amber mutant with a sequence CTAGG at the mutant site they obtained a reversion to a sense codon with the new sequence CCAGG. The new base sequence was now a hot spot with a factor 10 higher mutation rate than the unmethylated site. When the gene was transferred to E. coli B, which does not methylate the CCAGG sequences, the difference in mutation rate disappeared.

These experiments indicate that a change, presumably deamination of cytosine or 5-methylcytosine, which is known to occur in vitro at a rather high spontaneous rate, probably also occurs in vivo. Most cells have the enzyme uracil-DNA glycosidase which removes the uracil residue when the deamination of cytosine produces uracil. Additional repair enzymes then remove the lesion and replace a nucleotide containing cytosine before replication, and no lasting change is involved. However, if the base is 5-methylcytosine, the deamination product is thymine which will be like other thymine residues except that a mismatch T:G will be present. Enzymes that eliminate mismatched bases may repair the site, but since the G is as likely to be removed as the T, mutations are increased at the site by either repair replication or replication without repair. To test this effect the mutation rate of sites containing 5-methylcytosine were studied in a strain of E. coli with a Ung⁻ mutation, i.e., one which has the uracil-DNA glycosidase deleted.

Under these conditions the overall mutation rate was increased, but the difference between 5-methylcytosine and cytosine residues disappeared. These experiments indicate that in vivo deamination of 5-methylcytosine residues in DNA occurs at about the same frequency as the deamination of cytosine residues.

## 2. Are 5-Methylcytosine Residues Hotspots for Mutation in Eukaryotes?

Uracil-DNA glycosylase has been isolated from human cells (reviewed by LINDAHL 1982). It is a small protein similar to the *E. coli* enzyme, but has a twentyfold higher Km for dUMP residues in DNA. Nevertheless, one might expect 5-methylcytosine to be a hotspot for mutation in human and other mammalian cells and to behave much the same as these residues have been shown to do in bacteria. In line with this expectation, studies have shown that the frequency of CpG doublets has decreased in the course of evolution in many eukaryotic DNAs (RUSSELL et al. 1976) and the transition product, TpG and the complementary doublet CpA, have increased. In the SV 40 genome there are only 27 CpG doublets remaining in the 5,243 base pairs and nearly all of these are clustered around the origin of replication in the non-coding or regulatory regions of the genome. The TpG and CpA frequency in the genome is also correspondingly higher than expected from the base composition. Although both of these viral genomes are free of methylated cytosine in the virons and in the productive stage of viral infection (DIALA and HOFFMAN 1982, FORD et al. 1980), the loss of CpG doublets implies an earlier history of methylation, either in evolution or perhaps in the integrated state where mCpG sites would have been hotspots for transition to TpG and the complementary CpA in the other chain of the DNA helix.

Although the overall frequency of CpG doublets in many eukaryotic DNAs is lower than predicted by their overall base composition, some fractions of the genome have retained high levels of CpG doublets. Does that indicate that these fractions of the genome are not methylated and therefore are not hotspots for mutation from C:G to T:A? To try to answer we will consider a few specific examples. The bovine satellite I, a 1,400 bp tandem repeat that is present in about 100,000 copies per genome is one example that may be instructive. The CpG percentage is 6.4, almost that expected from the base composition and a random distribution of bases to produce CpG doublets. The variants revealed by sequencing the genomic DNA and several cloned segments of the 1,400 bp repeat show no particular pattern of mutation. Certainly the CpG sites are not hotspots for mutation in spite of the fact that nearly all of the cytosines in the CpG doublets are methylated in thymus DNA and most other somatic tissues (unpublished data). The sequence is the same in other tissues examined, including sperm DNA, but the methylation pattern is quite different in sperm where few if any CpG doublets are methylated. The methylation must occur in the somatic cells during differentiation of the early embryo, and the reason for the absence of hotspots may be that this satellite DNA is not methylated in the germline. Therefore, all mutants accumulated in somatic cells are lost each generation, while the germline transmits the DNA which has not been methylated and, therefore, is not subject to mutation by deamination of 5-methylcytosine.

Another example reveals a contrary situation in which sequences of the Alu dispersed repeats from human cells are hotspots for mutation. A high proportion of the variations seen when sequences are compared are restricted to the CpG sites as well as to the TpG and CpA doublets which in many instances arose from the mutation of a mCpG to TpC and CpA on the complementary chain. We may predict that the Alu sequences are methylated in the germline cells and have the resultant mutations transmitted to the offspring; hence the CpGs are hotspots for mutation in the Alu sequences. Perhaps this is a property of all sequences that are methylated in the germline. On the other hand, if a gene or repeated sequence is methylated only in the somatic cells, it will not drift rapidly and show variation at CpG sites in phylogenetic lines.

The 5 S RNA coding regions in eukaryotes have a rather high frequency of CpG doublets (ERDMAN 1982). The other ribosomal genes have also retained a relatively high level of CpG sites (SALIM and MADEN 1981). They have a variable level of methylation in somatic cells, but the extra-chromosomally amplified ribosomal coding regions are not methylated in the *Xenopus* oocyte (BIRD and SOUTHERN 1978). The methylation of ribosomal DNA has not been examined in other cells of the germline.

Another rather surprising example is the α-globin gene in human cells. The globin gene has an unexpectedly high frequency of CpG doublets in the coding regions (PROUDFOOT and MANIATIS 1980), but there is no evidence that these have been hotspots for mutation. The genes of the beta-globin cluster are highly methylated in the sperm DNA and the non-operating ones retain the methylation in the somatic cells. The α-globin genes may not yet have been examined. My prediction is that the α-globin gene is not methylated in germline cells. In this respect they may be characteristic of a larger family of genes that are unmethylated in the germline. They are non-functional because the other mechanisms necessary to turn on the genes after deletion of the methylation pattern and are not operating in the germline cells. Such genes might then be methylated during meiosis or in spermatids before maturation of sperm. Then differential deletion of the fully methylated pattern, which has been demonstrated by VAN DER PLOEG and FLAVELL (1980), would allow specific genes of the cluster to operate while the other genes of the cluster would remain suppressed in a particular reticulocyte. By contrast the human pseudogene (ψα1) has lost most of its

CpG sites presumably since it became non-functional. One could assume that release of the selection pressure has allowed it to drift freely, but that should have allowed the mutations to occur at random. However, when the sequence is compared with the α 2 globin gene (PROUDFOOT and MANIATUS 1980), most of the mutations are found at the CpG doublets. This implies that ψα1 became methylated as part of its inactivation and relegation to the status of a pseudogene. It was not only free to drift, but if methylated in the germline the mutations would have accumulated in the CpG doublets at a fast rate as one can see now by comparing the sequences.

## B. DNA Methylation and Repair

MESELSON (RADMAN et al. 1980) has proposed that methylation may be the distinguishing marker that allows the repair enzymes to recognize the new strand and repair it preferentially just back of the fork during replication. In this way proof reading enzymes, which operate following replication, could restore any mistakes which appear as base pair mismatches. After methylation of the newly replicated, hemimethylated DNA, this distinguishing feature would disappear. The hypothesis is not testable directly, but mutants at the *dam*-3 and *dam*-4 loci, which result in reduced levels of methylation at GATC sites, are found to have higher rates of mutation than control strains. They were more sensitive to base analogs which produce transitions that would appear as mismatches immediately after replication (GLICKMAN et al. 1978). Mutagenesis by an alkylating agent, EMS (ethyl methanesulfonate) was also enhanced in the mutant strains. However, the effect was much less when ultraviolet light was the mutagenic agent. Since the latter produces thymine dimers and few mismatches this observation supports the idea that the higher rates of mutation with the other agents are associated with mismatch repair.

A more direct test of the hypothesis was reported by RADMAN et al. (1980). In order to simulate the newly replicated DNA with one chain methylated and the new chain unmethylated, they produced heteroduplex lambda DNA with genetic markers suitable for following the mismatch repair. One DNA strand was grown in a mutant strain (dam$^-$) that had greatly reduced methylation of adenine in its DNA and the other lambda DNA strand was from a dam$^+$ strain. This duplex DNA was used to transfect wild type *E. coli*. In one experiment two well separated loci were used; c$^+$/c I and P$^+$/Pam-3 were the mismatches. The bias in the recovery of parental markers was about hundredfold favoring the methylated genotype. The recombinants c$^+$P$^+$ and cI Pam-3 arise by postreplicative recombination and by mismatch correction and are too complex for analysis here, but the corrections clearly favor the retention of the markers in the methylated chain. Correction did not occur if both chains of the transfecting DNA was methylated, but either hemimethylated DNA or unmethylated

DNA was repaired about equally. The repair was shown to involve distances of about 3,000 nucleotides that propagate in the 5′ to 3′ direction from the site of mismatches (WAGNER and MESELSON 1976).

These and subsequent experiments demonstrate a clear bias in the repair in which the nucleotide residue of the unmethylated chain is the one replaced. There is also an effect on recombination. KORBA and HAYS (1982) report experiments which indicate that hemimethylation at the CCAGG sites in *E. coli* DNA results in increased rates of recombination near the hemimethylated sites.

One of the problems in attributing to methylation the mechanism for strand recognition in mismatch repair at the growing fork in replication is the absence of methylation in some species. The very radiation resistant bacterial species *Micrococcus radiodurans* has no detectable methylation of its DNA and yeast has little if any methylation. The DNA of *Drosophila* and many other insects appears to be without methylation and yet one must suppose that those organisms are as efficient as *E. coli* in mismatch repair. There is another possible recognition signal for repair enzymes that appears to be universally characteristic of DNA replication. I refer to the presence of primer segments in the new chain preceding each Okazaki segment. In *E. coli* these segments are perhaps 1,000 nucleotides long, but in eukaryotes the segments are about 200 nucleotides. These RNA primers remain in the DNA for only a short time and then are excised and replaced with DNA. In the short time during which these RNA primers remain in DNA they could be an excellent marker for the new chain. The question might be raised about the leading strand which is replicated in the 3′ to 5′ direction and would not appear to require interruptions and new starts with RNA primer. HAY and DEPAMPILIS (1982) have presented evidence that the leading strand of SV 40 DNA is initiated within the origin by a mechanism similar to that which primes other Okazaki fragments, but the typical short fragments are not produced on the leading strand. However, even in this short genome additional starts on the leading strand are difficult to rule out. Many studies in the past have indicated that in eukaryotic systems the leading strand is also interrupted. The reason for the retention of RNA primers in DNA replication has never been understood. A compelling reason might be to favor new strand recognition for purposes of repair just back of the fork during replication. Unfortunately there does not appear to be a way to test the hypothesis, since mutants lacking RNA primers are extremely unlikely to be viable.

### C. How Stable Is a Pattern of Methylation?

If we assume that a pattern of methylation is preserved by maintenance methylase, we may ask what is known about the fidelity of the maintenance system. Unfortunately evidence on that point is scarce and the evaluation of the bits and pieces at hand is difficult. Would the maintenance of a pattern, based on an enzyme that scans the double helix for hemimethylated sites, methylates only these, and moves on over unmethylated sites, be as efficient as base pairing in replication? Information on this point is available from observations of what happens when methylated DNA is used to transfect or transform cells. As pointed out earlier there appears to be a significant loss of methylation before the integration of the DNA into the chromosome. Once that event has occurred the remaining methylated CpG doublets are maintained nearly as well as the sites in the regular genome (POLLACK *et al.* 1980). In studies with $\emptyset$ X 174 DNA STEIN *et al.* (1982) found that DNAs methylated in vitro with Hpa II methylase, which methylates both strands, lost 30–40% of their methyl moieties upon transfection. However, once established the remaining methylation pattern was stable for at least 100 cell generations. Infrequent losses of the order of mutation rates would not have been detected in these studies. However, the techniques for measuring retention, restriction enzyme digestion, are accurate enough to detect a more efficient retention of the pattern when hemimethylated $\emptyset$ X 174 DNA rather than enzymatically modified DNA was used to transform mouse L-cells. The hemimethylated DNA was prepared by primer repair synthesis using single stranded $\emptyset$ X 174 DNA as a template. The reaction mixture contained 5-methyl-deoxycytidine 5'-triphosphate and the other three nucleoside triphosphates. The transforming DNA then had essentially all cytosines substituted with 5-methylcytosine. Only the CpG sites maintained the methyl cytosine during replications, but the retention in transformed cells was better than with Hpa II methylated DNA and nearly complete after integration. The loss of methylation before or during integration is presumably happening by replication without efficient maintenance methylation, but once integrated the remaining sites are then efficiently maintained. Could the hemimethylated state, or the complete methylation of all cytosines in the DNA, have delayed its replication and thereby prevented the dilution of the pattern until integrated? The data are not available to give an answer, but the proposal that hemimethylated DNA would not replicate again until fully methylated (TAYLOR 1977) might be an explanation. If one supposes the hemimethylated DNA is methylated only inefficiently without replication (HARLAND 1982), perhaps by a repair type methylase, or when replicated without the functioning of a homologous origin, the observed results could be explained.

WIGLER *et al.* (1981) studied the retention of methylated Hpa II sites in the chicken thymidine kinase gene that had been cloned into pBr 322 and methylated in vitro with Hpa II methylase. The methylation decreased its apparent transformation efficiency, but once integrated the methylated genomes were suppressed in expression of the kinase gene. The methylation pattern was retained, but with less than 100% efficiency. They calculated the retention to be $94 \mp 3\%$ effective at HpaII sites each cell cycle when measured about 25 cycles after transformation. However, the calculation was based on one determination at the end of the 25 cycles. No evaluation could be made of the changes which might have occurred during the transformation events as distinguished from those after transformation. On the same basis the SmaI site, CCCGGG, appeared to be less efficiently maintained, about 85% per cycle. The HpaII methylation of $\varnothing$X174 was maintained with an efficiency of 95%. All of these frequencies are too inefficient to be a useful genetic mechanism. It is possible that most of the loss occurred either during the integration or even before integration if the DNA was replicated before integration. Perhaps there are variations depending on the site of integration and the maintenance of methylation on genes in their normal position in the genome is a better indicator of the fidelity of the maintenance process.

HARRIS (1982) studied the rate of reversion of $TK^-$ mutants in a clonal isolate (204–7) from V79 Chinese hamster cells. The mutant had been selected from isolates produced by long-term culture in BrdU (bromodeoxyuridine) under light that kills cells which incorporate the BrdU into their DNA. These selected mutants survive because they lack a detectable amount of thymidine kinase activity necessary for the phosphorylation of BrdU and its subsequent incorporation into DNA. The mutant was very stable with a reversion rate of approximately $10^{-7}$ and persists in culture in drug-free medium. However, 24-hour treatment with 5-azacytidine causes high rates of reversion to TK. At a concentration of 1 $\mu$M the reversion was greater than $10^{-3}$. At this concentration cell survival is about 30%.

JONES and TAYLOR (1980, 1981) have shown that azacytidine under these conditions caused a significant deletion of DNA methylation which persists in the surviving cells. Therefore, it is probably safe to assume that the reversion from $TK^-$ to $TK^+$ in untreated cells is a measure of the spontaneous rate or fidelity of maintenance of CpG methylation. The unknown paramter is the number of CpG sites that are involved in the suppression of the TK gene in the Chinese hamster. The gene will have to be cloned, sequenced and the methylation patterns determined before one can evaluate the data in terms of maintenance of the pattern. The origin of the mutants suggests that several sites may be involved and the loss of one is unlikely to allow expression. The mutant 204–7 was isolated after a long period of selection in BrdU where TK activity would have resulted in death. The first stage in the selection from wild-type cells is the production of partially resistant clones that can be isolated by one-step selection in BrdU. These cells have reduced levels of thymidine kinase, but the residual is sufficient for them to grow in HAT medium (hypoxanthine, aminopterin and thymidine). After serial culture for very many cell generations the resistant cells with the low reversion rates were finally isolated. This suggests steps with different levels of methylation before the highly resistant cells are obtained. The variations in TK activity in revertants also suggests several sites are involved with some perhaps more efficient for suppression than others. Thymidine kinase activity ranged from 0.3–0.5 percent in mutant 204–7 and in two other TK$^-$ mutants (RJK-92 and B150-7) to 7 and 10% of the amount in the parental line V 79-56. Several lines of mouse cells that were TK$^-$ failed to revert with azacytidine and therefore probably have a change in the gene that is not related to methylation. Several other types of data also indicate that maintenance of methylation causes the TK$^-$ state in hamster. For example, several known mutagens, ethylmethane sulfonate, nitrosoguanidine and ultraviolet light, increased the rate over background only three- to eightfold. Other agents known to interfere with methylation (Na butyrate, L-ethionine and 3-deazadenosine) greatly increased the number of HAT resistant colonies. These latter agents do not significantly increase mutation rates of other genes.

HARRIS' experiments indicate that maintenance of methylation, if that is the mechanism of suppression of the TK gene in Chinese hamster cells, can be very effective. Since two or more sites may be involved in each activation event, the fidelity of maintenance can not be deduced directly from these experiments even if methylation is the only change. The failure of any of the revertants to yield the amount of TK activity of the parental V 79 means either that deletion of methylation is incomplete or some other mutation in the gene occurred before or during its methylation.

## D. Overview on the Role of DNA Methylation

The discovery of heritable modifications of DNA that are reversible in a stepwise fashion has changed our perspectives in considering differentiation. Nearly a century ago Weisman proposed that the genetic units were segregated to the cells in such a way as to restrict their potential in development. Embryologists and geneticists began testing these ideas in a variety of ways. The embryologists developed the concept of the totipotent cell and the experiments with nuclear transplants in animals and development of whole plants from single cells appeared to rule out such segregation or loss of genetic units except in a few isolated cases. Genetic and cytogenetic studies also established the integrity of the genome and its transmission to all of the body cells again with a few exceptions in which the germline nevertheless carries those chromosomes or parts of chromosome that are eliminated in the somatic cell lineage. However, with the demonstration that genes can be preserved in an inactive state by mobile genetic elements or by methylation which can be inherited with high fidelity, we come full circle. The potential may be limited, not by loss, but by modifications of certain genes in an inactive or partially active state. Of course, these concepts, even if accepted, do not allow us to answer the most fundamental questions about how these changes are controlled in a stepwise fashion.

Other less stable changes in the genome mediated by the binding of other molecules to DNA that modify its three-dimensional structure without changing the one-dimensional coding sequences may also play a role in differentiation. However, there is no convincing demonstration that these changes can be transmitted through numerous cell divisions without an accompanying modification of the bases or their sequence. The modifications, that are inducible by regulatory genes and can be locked into perpetuating systems, are candidates for these three-dimensional changes (MONOD and JACOB 1961). In higher organisms such changes include those initiated by steriod hormones and their receptor proteins and by cell surface acting hormones and other proteins. The change to Z form DNA may also be in this class, but it can be related to changes in methylation and possible mobile genetic elements. It is also difficult to classify the changes involved in making the functional genes or their flanking sequences more susceptible to nucleases compared to the same genes in other tissues where they are inactive. Some of these modifications have been correlated with changes in methylation, but others seem not to be. Likewise, the transition points between Z form and B form DNA are more susceptible to single stranded nucleases.

The correlations between altered nucleosome structure and increased sensitivity to nucleases has been recently considered in an interesting and thoughtful review by WEISBROD (1982). He emphasizes the role of modified chromatin proteins and the switches that have been demonstrated in the histones and the HMG (high mobility group) proteins. However, the same problem that has always challenged us without an adequate answer remains. How can protein binding achieve differentiation without alteration of the DNAs primary structure. One suspects that the two events might be correlated. Which is cause and which effect? Let us see where our present knowledge of methylation can lead.

There are two classes of genes that appear to operate differently with respect to methylation. A large group of genes which function in somatic cells are highly methylated in sperm DNA. These presumably arrive in the zygote in a methylated and inactive state. Their activation requires replication and cell division to "uncover" the genes in a stepwise fashion. Another group of genes or DNA sequences, including some of the highly repeated DNA, arrive in the zygote in an unmethylated state. These DNA sequences may become methylated during development. Included in this group are certain integrated viral genomes (JAHNER et al. 1982) and if we take a clue from their experiments the unmethylated genes are methylated in the early embryo, before implantation in the mammal. If they do not become methylated at that stage, methylation is unlikely and usually does not occur during the life of that animal. What could be the role of a group of genes that are unmethylated in the germline and then become methylated early in development? The simple answer might be that these sequences include genes that are required in the germline cells, but have no necessary function in the soma. If that is so why are retroviral genomes in that category along with certain tandemly repeated sequences? My guess is that once a category of genes had evolved with a mechanism for inactivation in the soma, any other group could be added if their methylation signal came under the same control system. If the signal resides in a mobile genetic element similar to the human Alu sequence, various genes could over time acquire a similar pattern regardless of their function. As long as their inactivation conserved energy, which their transcription would expend, and the transcripts are not required such groups would tend to accumulate during evolution. Cells would have then acquired a relatively simple mechanism for handling excess genetic baggage.

Since our information suggests that the differentiated state of a cell changes only when its DNA is replicated, the dispersed repetitive sequences that signal a change in methylation pattern, whether through the action of a de novo methyl transferase or an inhibitor of the maintenance methylase which uncovers the gene, should be associated with origins of replication. The tentative selection of Alu equivalent repeats for the role has the advantage of the proper correlation of functions, but there is no experimental evidence for or against the idea yet available. In principle such evidence can now be obtained by transforming with genes linked to different Alu type repeats. For example, the Alu sequences flanking the gamma-globin genes on the 5' side in the human cluster could be replaced by the one on the 5' side of the delta and beta-globin genes in a vector that could be used in transformation or for injection into the zygote. The gene would be hemimethylated before transformation according to the technique used by FLAVELL et al. (1982). One would then see if the methylated gene would be uncovered, i.e., lose its methylation pattern and be turned on simultaneously with the adult genes rather than the fetal globin genes. Of course, one would not use the human genes but comparable animal genes, which could be manipulated in the experimental animal and their function followed in vivo during development. The mouse hemoglobin genes, if they turn out to have dispersed repeated sequences appropriately placed, could be used to transform early embryos which would be returned to the uterus for development. A more convenient experimental system might be the Xenopus globin genes if these have appropriately located dispersed repeats flanking the hemoglobin genes.

Another system that may offer a more direct approach would be the retroviral genomes which become methylated in the early embryo. If one could insert into the viral genome a gene known to be inactivated by methylation the site active in the control of de novo methylation might be located. For example, the well characterized herpes virus TK gene (McKNIGHT and GARVIS 1980, McKNIGHT and KINGSBURY 1982) is inhibited when methylated at CpG sites (CHRISTY and SCANGOS 1982). If it were inserted in the retroviral genome, that can be integrated and methylated in early mouse embryos before implantation (JAHNER et al. 1982), the effect on the TK gene could be followed by its expression. If it also becomes methylated then systematic studies in which parts of the retroviral genome was used to transform might give clues about the regulatory region.

### E. An Hypothesis for the Control of Methylation Patterns

A discussion of the control of the methylation patterns in the human hemoglobin genes by proximity of Alu sequences 5′ to the coding region was initiated in Section III. We will develop that idea further. Let us assume that the DNA methylases have a specificity for the CpG sites and no requirement for a particular nearby flanking sequence. We will make the further assumption that maintenance methylases which operate immediately following replication bind only at RNA-DNA hybrid sites (primer sites) and move along the DNA in the 5′ to 3′ direction of the new unmethylated DNA chains. Every CpG site which is hemimethylated will be methylated at the symmetrical site on the new chain while any CpG without one chain already methylated will be passed over. The binding of certain inhibitor molecules to the primer site on the leading strand will prevent the methylase from binding and prevent methylation of the whole region downstream to the next binding site. If these binding sites on the leading strand are the Alu type of dispersed repeat, there could be enough variation and specificity to limit a specific inhibitor to a family of genes. The Alu sequence on the 5′ flank of the human gamma-globin genes, for example, is different enough (80% homology according to DUNCAN et al. 1981) from the Alu sequence on the 5′ side of the delta and beta genes to be distinguished by a DNA binding protein. Perhaps there are Alu sequences flanking other genes that are turned on simultaneously with the human gamma gene. Since none are yet known we will dismiss this idea for the present except to suggest that it would be extremely unlikely that very large numbers of specific inhibitors would have evolved if one could serve to block the methylase binding sites of many genes that are needed at one stage in development. The inhibitor remains in place during one or two cell cycles during which a new methylation pattern is set up. The methylation would be effectively deleted as soon as a hemimethylated DNA segment was replicated a second time (Fig. 4). The CpGs involved could be all of those between two successive binding sites, i.e., between two functional Alu sites in this model. Other dispersed repeats may also be candidates for binding sites, but to avoid confusion we will think only in terms of Alu sequences for the purpose of this model.

The specific inhibitors then become the key molecules in differentiation. How could the appearance of these molecules be regulated during development and differentiation? There is no answer except to assume that a programmed activation of their coding sequences operates in a cascade rather early in development. The first group of inhibitors activated by fertilization operate for a short time and then induce the next group which can operate only after the first eschelon has produced their effect. These stepwise rescues of genes from the inactive state would be crucial to normal development, but would not necessarily turn on the genes. Changes in methylation patterns would represent the stage of determination. Differentiation would have to involve additional steps to activate transcription and the subsequent steps in protein synthesis, transport and utilization of the final products.

The changes in methylation which come late in development such as those which occur in the lymphocytes to activate J chain synthesis and the appropriate constant heavy chain gene (Section III) pose a special problem. Since an embryonic cascade could not last long enough for these situations, some other mechanism would have to be envisioned. The suggestion by OTT et al. (1982) that transcription may result in the loss of methylation at the functioning gene does not solve the problem. One has to think of a way in which transcription can override the negative control of methylation. Once that has happened the transcription or some co-product of transcription might inhibit the maintenance methylase. The stimulus to transcription, a hormone and its receptor complex, for example, might operate on a partially active gene and prepare it for a greater activity from a future stimulation as suggested by OTT et al. (1982).

There should be at least three classes of genes: (1) those that are modulated by methylation, either completely inhibited or reduced in activity, (2) those unaffected by methylation either because they are not methylated or because the modifications do not affect affinity for molecules involved in their transcription, and (3) those that require methylation to operate. All three types have been tentatively identified although we have placed our attention on the first class. The second class would include those genes that are required to operate in all types of cells, presumably those coding for enzymes of intermediate metabolism are some examples.

Examples of the first class would be the hemoglobin genes, where conditions for the expression of any of the genes of the cluster may be present in the precursor cells of erythrocytes. The ones actually expressed are the transcribed ones, and differential methylation is a model system for this situation. Most of the other examples cited in this review are of this type, but the vitellogenin genes of *Xenopus* may fall into class 2 (GERBER-

HUBER *et al.* 1983). However, more extensive mapping of the methylated CpG sites not a part of HpaII and HhaI sites is necessary before that example can be verified. Histone genes might also be in class 2, since they are used in all cells. However, there are some histone genes that are used only in a particular cell lineage or some special situation and these could utilize a differential methylation as part of their regulatory system.

Another example of genes that have eluded the control by methylation are the early genes of SV 40 and polyoma virus. SV 40 has done this by elimination of nearly all of the CpG sites in the early gene region primarily by mutation to TpG sites. Only one near the 5′ terminus of the late transcript and two within remain (FRADIN *et al.* 1982). Polyoma has a few more but neither of these viruses is inhibited in productive infection by the methylation of sites in the whole genome (SIMON, personal communication). On the other hand some of the remaining CpG sites inhibit transcription of the late region when methylated.

An example of the requirement for methylation to function should also be a possibility with the tinkering that characterizes evolution (JACOB 1981). HATTMAN (1982) may have discovered one of this class in bacteria where a phage Mu gene (*mom*) controls an unusual modification of the phage DNA. This modification requires the host *dam* gene which codes for a DNA adenine methylase. In *dam*⁻ hosts the Mu DNA is not modified. Probes were prepared by nick-translation of either the whole viral genome or a restriction fragment containing the *mom* gene and hybrized with RNAs transferred to nitrocellulose filters after separation by electrophoresis. He found (1) non-specific RNA in *dam*⁺ lysogenic cells, but only after induction of the Mu prophage, (2) there was only about one twentieth as much *mom* RNA in *dam*⁻ Mu lysogens and (3) the same amount of total Mu transcripts in both *mom*⁺ and *mon*⁻ cells when probed with the whole Mu genome. These experiments indicate that transcription of the *mom* gene requires an active DNA methylase gene, the product of the *dam* gene presumably, to methylate the gene at GATC sites.

Probably some cells in the adult maintain the embryonic methylation pattern or the major part of it. The teratocarcinoma cell whould be an example that has proliferated and formed a teratoma. Other tissues may contain a few such cells and this could account for the rare successes with transplanting nuclei from adult cells back into the egg of the frog. Higher plants may have even more such cells. Instead of having a germline, they might retain dispersed embryonic cells which, with the proper stimulation and environment, can regenerate the whole plant. As one moves up the phylogenetic scale of animals and finds that regeneration of parts is more and more limited, the frequency of the embryonic cells may decrease to the vanishing point in some tissues.

Certainly we can not satisfactorially answer all questions about differentiation, but methylation is a modification of the genome that can be transmitted with high fidelity. It would indeed be surprising if some use has not been made of it in differentiation. Just how often and how significant the use has been remains to be demonstrated. Methylation may serve other roles, and indeed the evidence suggests that not all of it is likely to be involved in the control of transcription. The abundant 5-methyl cytosine in higher plants is still a puzzle. The expansion of the sites to include CpXpG sites in addition CpG sites explains how these cells can accommodate so much of the methylated base (GRUENBAUM et al. 1981) but offers no clue to its functional significance.

## F. A Pyramid of Controls in Vertebrate Cells

Several observations raise doubts about the universality of methylation as the major regulatory mechanism. The first and most disturbing is that insects, in particular *Drosophila,* do not have the CpG methylation which appears to be effective in suppression of gene transcription in vertebrate cells. The second are instances such as that reported by GERBER-HUBER et al. (1983) for the vitellogenin genes of *Xenopus* in which no changes in methylation pattern can be detected. The vitellogenin genes are not transcribed in liver cells of the male *Xenopus* unless the animals are injected with estrogen. The genes are naturally induced in females when yolk formation in eggs is occurring. No differneces between methylation pattern in males and females can be detected in the genes or their flanking regions before or during the induction by hormone injection. A third instance is the failure of WOLF and MIGEON (1982) to detect any consistent differences in the methylation of active and inactive X chromosomal DNA in human cells. An explanation may be that methylation is indeed not the universal regulatory mechanism, but a second level in a pyramid of controls.

The top level could be the one connected with the time of replication of the DNA over the cell cycle. DNA appears to be replicated in either the first half of S phase, $S_E$, or the last half of S phase, $S_L$ (KLEVECZ and KENISTON, 1975; DUTRILLAUX et al., 1976). The DNA replicated in $S_E$ contains the active genes while that replicated in $S_L$ contains the inactive genes. This was first indicated by experiments of KAJIWARA and MUELLER (1964) who reported that bromodeoxyuridine (BrdU) incorporated into DNA in $S_E$ causes a marked reduction in cloning efficiency. However, the same amount of incorporation in $S_L$, has no effect (LOUGH and BISCHOFF, 1976). In Friend murine erythroleukemia cells, BrdU causes a delay in induction of hemoglobin synthesis only if incorporated into the DNA during (BROWN and SCHILDKRAUT, 1979). Later reports showed that the globin genes are replicated in $S_E$ in erythroeleukemic cells where the adult globin genes are inducible with dimethyl sulfoxide and certain other drugs (EPNER et al., 1981; FURST et al., 1981). Recently GERALD HOLMQUIST, Baylor College of

Medicine, and colleagues (personal communication) have shown that the hemoglobin genes are replicated in $S_L$ in human fibroblasts where the genes are not inducible.

The inactive X chromosome has been known to be late replicating for many years, but was usually considered a special situation. It may be in the sense that the genes on only that chromosome are shifted to replicated in $S_L$, but otherwise it may be typical of many inactive genes—they are replicated in late S phase. This implies that many genes are in replicons that can be programmed to replicate either in $S_E$ or $S_L$ depending on a modification of the origin of replication. This control system would be at the top of the pyramid. If a gene sequence is replicated in $S_E$, it probably binds histone and certain non-histone proteins including the HMG proteins (reviewed by WEISBROD, 1982). These make the sequences susceptible to nucleases at certain sites and represent the first step in making the genes available to transcription. For some genes that first step could be sufficient, while for others, subsequent modificatons are required. The types of proteins bound by the newly replicated DNA would be determined by the types of proteins available in $S_E$. DNA synthesized in $S_L$ would be complexed with histones and perhaps other proteins that keep them from being inducible.

The next level in the pyramid would be differential DNA methylation of sequences, a change that deletes methylation at certain sites or leads to *de novo* methylation at unmethylated sites. The first level, time of replication, would be under the control of a master origin of replication in a replicon that contains a cluster of genes and suborigins (Alu-type dispersed repeats) as proposed by TAYLOR (1983). The function of the suborigins may be to control replication of the DNA in their domain, but more important for our discussion, these repeats would be sites for binding of methylases and inhibitors of methylases. These transcribable repeats are often located between each gene or in some cases between groups of two or more coordinately functional genes, as pointed out for the cluster of β-like hemoglobin loci in the human genome. The repeats vary enough in base sequence to store information of the type proposed, but only some very carefully planned experiments can reveal such differences. If a replicon is switched to $S_E$ during the determination steps in early development, the next step might be deletion of the methylaton pattern of those genes which can later be transcribed. If we use the globin genes in the human β-like cluster as an example, the stem cells that become precursors for adult red cells would have the methylation changed or deleted in the β and δ region by the binding of a methylase inhibitor at the 5' flanking Alu repeat. The inhibitor would have to remain attached though two cycles to delete methylation from both chains unless the demethylating activity discovered by GJERSET anf MARTIN (1982) is a hemi-demethylase. If the enzyme removes the methyl groups from only one strand and operates in conjunction with a methylase inhibitor, the deletion could be accomplished in one cell cycle. The demethylating activity was reported by GJERSET and MARTIN to remove only about one half of the methyl groups which is consistent with a hemimethylase activity. Subsequent steps in the regulatory pyramid are required for function even when the erythroleukemia cell has the globin genes replicated in $S_E$ and the β-globin gene is in a functional state with respect to its methylation pattern. Therefore, the pyramid may have several steps, but the top two are time of replication and changes in methylation pattern for some genes. For other genes, vitellogenin genes, the methylation step is not part of the control. Likewise, in *Drosophila* that level of regulation may have been deleted for all of the genes.

One question remains, if the above hypothesis is correct. How are the master origins modified to switch from $S_E$ to $S_L$ or the reverse. Since the change is inherited we may propose a change in methylation of the origin. However, since CpG methylation regulates transcription perhaps the origin methylation would be the rarer CpC methylation in which the modified site is $^mCpC$ or some other sequence not yet detected. A second possibility is the action of mobile genetic elements at origins. Since most Alu repeats are bracketed by short direct repeats (SCHMID and JELINEK, 1982), the suggestion that they may have moved to their present location has been made. The master origins might be peculiar doublets of the single Alu sequences (TAYLOR, 1983). A third possibility but very unlikely, is some type of restricted mutation similar to that suggested by SCARANO et al. (1977). For example, if the sites contain 5-methyl cytosine in $^mC : G$ base pairs, the deamination of cytosine would produce a $T : A$ pair after replication or repair. The reverse change could involve the methylation of adenine and oxidative deamination at the methyl amine group to produce hypoxanthine which would pair with cytosine and restore the $C : G$ base pair. Such enzymatically directed mutations are of course, very unlikely.

The only evidence that indicates the type of change that might be involved in the modification of master origins are the experiments that show that azacytidine treatment reactivates genes on the inactive X chromosome (MOHANDAS et al., 1981). It is, of course, possible that the methylation pattern of the genes has been changed, but it is just as likely that the time of replication of blocks of genes in the inactive X are switched from $S_L$ to $S_E$ and that the change of methylation in or near the affected gene if it occurs is only one part of the activation. Both a change in the origin and a change in CpG methylation in the 5′ flanking region could be required for function.

# References

ADAMS, R. L. P., 1971: The relationship between synthesis and methylation of DNA in mouse fibroblasts. Biochim. Biophys. Acta **254**, 205—212.

— 1974: Newly synthesized DNA is not methylated. Biochim. Biophys. Acta **335**, 365—373.

ADAMS, R. L. P., BURDON, R. H., 1982: DNA methylation in eukaryotes. Critical Reviews in Biochem. **13**, 349—384.

ARBER, W., 1965: Host-controlled modification of bacteriophage. Ann. Rev. Microbiol. **19**, 365—378.

— DUSSOIX, D., 1962: Host specificity of DNA produced by *Escherichia coli*. I. Host controlled modification of bacteriophage. J. Mol. Biol. **5**, 18—36.

— LINN, S., 1969: DNA modification and restriction. Ann. Rev. Biochem. **38**, 467—500.

ASTRIN, S., 1978: Endogenous viral genes of the White Leghorn chicken: Common site of residence and sites associated with specific phenotypes of viral gene expression. Proc. Natl. Acad. Sci. U.S.A. **75**, 5941—5945.

BACHI, B., REISER, J., PIRROTTA, V., 1979: Methylation and cleavage sequences of the Eco P 1 restriction-modification enzyme. J. Mol. Biol. **128**, 143—163.

BARALLE, F. E., SHOULDERS, C. C., GOODBOURN, S., JEFFREYS, A., PROUDFOOT, N. J., 1980: The 5' flanking region of human epsilon-globin gene. Nucl. Acids Res. **8**, 4393—4404.

BARR, M. L., BERTRAM, E. G., 1949: A morphological distinction between neurons of the male and female, and the behaviour of the nucleolar satellite during accelerated nucleoprotein synthesis. Nature **163**, 676—677.

BEACH, L. R., PALMITER, R. D., 1981: Amplification of the metallothionein-I gene in cadmium-resistant mouse cells. Proc. Natl. Acad. Sci. U.S.A. **78**, 2110—2114.

BEERMAN, W., 1952: Chromomeren-Konstanz und spezifische Modifikationen der Chromosomenstruktur in der Entwicklung und Organdifferenzierung von *Chironomus tentans*. Chromosoma (Berl.) **5**, 139—198.

— 1962: Riesenchromosomen. (Protoplasmatologia/Handbuch der Protoplasmaforschung, Band IV d.) Wien: Springer.

BEHE, M., ZIMMERMAN, S., FELSENFELD, G., 1982: Changes in the helical repeat of poly(dG-m5dC) · poly(G-m5dC) and poly(dG-dC) · poly(dG-dC) associated with B-Z transition. Nature **293**, 233—235.

BELL, G. I., PICTET, R., RUTTER, W. J., 1980: Analysis of the regions flanking the human insulin gene and sequence of an Alu family member. Nucl. Acids Res. **8**, 4091—4109.

BENNETT, P. M., GRINSTED, J., RICHMOND, M. H., 1977: Transposition of TnA does not generate deletions. Mol. Gen. Genetics **154**, 205—211.

BENZER, S., 1961: On the topography of the genetic fine structure. Proc. Natl. Acad. Sci. U.S.A. **47**, 403—416.

BESSMAN, M. J., LEHMAN, I. R., ADLER, J., ZIMMERMAN, S. B., SIMMS, E. S., KORNBERG, A., 1958: Enzymatic synthesis of deoxyribonucleic acid. III. The incorporation of pyrimidine and purine analogs into deoxyribonucleic acid. Proc. Natl. Acad. Sci. U.S.A. **44**, 633—640.

BIRD, A. P., SOUTHERN, E. M., 1978: Use of restriction enzymes to study eukaryotic DNA methylation: I. The methylation pattern in ribosomal DNA from *Xenopus laevis*. J. Mol. Biol. **118**, 27—47.

— TAGGART, M. H., GEHRING, C. A., 1981 a: Methylated and unmethylated ribosomal RNA genes in the mouse. J. Mol. Biol. **152**, 1—17.

— — MACLEOD, D., 1981 b: Loss of rDNA methylation accompanies the onset of ribosomal gene activity in early development of *X. laevis*. Cell **26**, 381—390.

BOLEN, P. L., GRANT, D. M., SWINTON, D., BOYNTON, J. E., GILLHAM, N. W., 1982: Extensive methylation of chloroplast DNA by a nuclear gene mutation does not affect chloroplast gene transmission in *Chlamydomonas*. Cell **28**, 335—343.

BOREK, E., 1963: The methylation of transfer RNA. Cold Spr. Harb. Symp. Quant. Biol. **28**, 139—148.

— RYAN, A., ROCKENBACH, J., 1964: Nucleic acid metabolism in starving bacteria. Fed. Proc. **13**, 578.

BRIGGS, R., KING, T. J., 1952: Transplantation of living nuclei from blastula cells into enucleated frog's eggs. Proc. Natl. Acad. Sci. U.S.A. **38**, 455—463.

BROWN, E. H., SCHILDKRAUT, C. L., 1979: Perturbation of growth and differentiation of Friend murine erythroleukemia cells by bromodeoxyuridine incorporation in early S phase. J. Cell Physiol. **99**, 261—278.

BURR, B., BURR, F. S., 1982: *Ds* controlling elements of maize at the *shrunken* locus are large and dissimilar insertions. Cell **29**, 977—986.

CAMERON, J., LOH, E. Y., DAVIS, R. W., 1979: Evidence for transposition of dispersed repetitive DNA families in yeast. Cell **16**, 739—751.

CATTANACH, B. M., 1962: XO mice. Genet. Res. **3**, 487—490.

CHALEFF, D. T., FINK, G. R., 1980: Genetic events associated with the insertion mutation in yeast. Cell **21**, 227—237.

CHAMBERS, J. C., TAYLOR, J. H., 1982: Induction of sister chromatid exchanges by 5-fluorodeoxycytidine: correlation with DNA methylation. Chromosoma **85**, 603—609.

— WATANABE, S., TAYLOR, J. H., 1982: Dissection of a replication origin of *Xenopus* DNA. Proc. Natl. Acad. Sci. U.S.A. **79**, 5572—5576.

CHRISTMAN, J. K., PRICE, P., PEDRINAN, L., ACS, G., 1977: Correlation between hypomethylation of DNA and expression of globin genes in Friend erythroleukemia cells. Eur. J. Biochem. **81**, 53—61.

— WEICH, N., SCHOENBRUN, B., SCHNEIDERMAN, N., ACS, G., 1980: Hypomethylation of DNA during differentiation of Friend erythroleukemia cells. J. Cell Biol. **86**, 366—370.

CHRISTY, B., SCANDOS, G., 1982: Expression of transferred thymidine kinase genes is controlled by methylation. Proc. Natl. Acad. Sci. U.S.A. **79**, 6299—6303.

COHEN, J. C., 1980: Methylation of milk-borne and genetically transmitted mouse mammary tumor virus proviral DNA. Cell **19**, 653—662.

COLLINS, M., RUBIN, G. M., 1982: Structure of the *Drosophila* mutable alleles, white-crimson and its white-ivory and wild-type derivatives. Cell **30**, 71—79.

COMPERE, S. J., PALMITER, R. D., 1981: DNA methylation controls the inducibility of the mouse metallothionein-I gene in lymphoid cells. Cell **25**, 233—240.

COULONDRE, C., MILLER, J. H., FARABAUGH, P. J., GILBERT, W., 1978: Molecular basis of base substitution hotspots in *Escherichia coli*. Nature **274**, 775—780.

CURRY, J. L., TRENTIN, J. J., 1967: Hemopoietic spleen colony studies: I. Growth and differentiation. Dev. Biol. **15**, 395—413.

DANOS, O., KATINKA, M., YANIV, M., 1980: Molecular cloning, refined, physical map and heterogeneity of methylation sites of papilloma virus type 1 a DNA. Eur. J. Biochem. **109**, 457—461.

DAVIDSON, R. G., NITOWSKI, H. M., CHILDS, B., 1963: Demonstration of two populations of cells in the human female heterozygous for glucose-6-phosphate dehydrogenase variants. Proc. Natl. Acad. Sci. U.S.A. **50**, 481—485.

DE LUCIA, A. L., LEWTON, B. A., TJIAN, R., TEGTMEYER, P., 1983: Topography of Simian Virus 40 A protein-DNA complexes: Arrangement of pentanucleotide interaction sites at the origin of replication. Virol. **46**, 143—150.

DESROSIERS, R. C., MULDER, C., FLECKENSTEIN, B., 1979: Methylation of *Herpesvirus saimiri* DNA in lymphoid tumor cell lines. Proc. Natl. Acad. Sci. U.S.A. **76**, 3839—3843.

DIALA, E. S., HOFFMAN, R. M., 1982: Hypomethylation of HeLa DNA and the absence of 5-methylcytosine in SV40 and adenovirus (type 2) DNA: Analysis by HPLC. Nucl. Acids Res. **107**, 19—26.

DIBERARDINO, MARIE E., HOFFNER, NANCY J., 1983: Gene reactivation in erythrocytes: Nuclear transplantation in oocytes and eggs of *Rana*. Science **219**, 862—864.

DOERFLER, W., 1981: DNA methylation—a regulatory signal in eukaryotic gene expression. J. Gen. Virol. **57**, 1—20.

DOSKOCIL, J., SORM, F., 1962: Distribution of 5-methylcytosine in pyrimidine sequences of deoxyribonucleic acids. Biochim. Biophys. Acta **55**, 953—959.

DRAHOVSKY, D., LACKO, I., WACKER, A., 1976: Enzymatic DNA methylation during repair synthesis in non-proliferating human peripheral lymphocytes. Biochim. Biophys. Acta **447**, 139—143.

— MORRIS, R. N., 1971 a: Mechanism of action of rat liver DNA methylase. I. Interaction with double stranded methyl-acceptor DNA. J. Mol. Biol. **57**, 475—489.

— — 1971 b: Mechanism of action of rat liver DNA methylase. II. Interaction of single-stranded methyl-acceptor DNA. J. Mol. Biol. **61**, 343—356.

— WACKER, A., 1975: Enzymatic methylation of replication-DNA intermediates in Ehrlich ascites tumor. Naturwissenschaften **62**, 189—190.

DUNCAN, C. H., JAGADEESWARAN, P., WANG, R. R. C., WEISSMAN, S. M., 1981: Structural analysis of templates of RNA polymerase III transcripts of Alu family sequences interspersed among the human beta-like globin genes. Gene **13**, 185—196.

DUNN, D. B., SMITH, J. D., 1955: Occurrence of a new base in the deoxyribonucleic acid of a strain of *Bacterium coli*. Nature **175**, 336—337.

DUTRILLAUX, B., COUTURIER, J., RICHTER, C.-L., VIEGAS-PEQUIGNOT, E., 1976: Sequence of DNA replication in 277 R- and Q-bands of human chromosomes using BrdU treatment. Chromosoma **58**, 51—61.

DYER, T. A., 1982: Methylation of chloroplast DNA in *Chlamydomonas*. Nature **298**, 422—423.

EHRLICH, M., EHRLICH, K., MAYO, J. A., 1975: Unusual properties of the DNA from *Xanthomonas* phage XP-12 in which 5-methylcytosine completely replaces cytosine. Biochim. Biophys. Acta **395**, 109—119.

EPNER, E., RIFKIND., R. A., MARKS, P. A., 1981: Replication of $\alpha$- and $\beta$-globin DNA sequences occurs during early S phase in murine erythroleukemia cells. Proc. Natl. Acad. Sci. **78**, 3058—3062.

ERDMAN, V. A., 1982: Collection of published 5 S and 5.83 RNA sequences and their precursors. Nucl. Acids Res. **10**, r93—r115.

ERREDE, B., CARDILLO, T. S., WEVER, G., SHERMAN, F., 1981: Studies of transposable elements in yeast. I. ROAM mutations causing increased expression of yeast genes: Their activation by signals directed toward conjugation functions and their formation by insertion of Tyl repetitive elements. Cold Spr. Harb. Symp. Quant. Biol. **45**, 593—602.

FELBER, B. K., GERBER-HUBER, S., MEIER, C., MAY, F. E. B., WESTLEY, B., WEBER, R., RYFFEL, G. V., 1981: Quantitation of DNase I sensitivity in *Xenopus* chromatin containing active and inactive globin, albumin and vitellogenin genes. Nucl. Acids Res. **9**, 2455—2474.

FENG, T. Y., TU, J., KUO, T. T., 1978: Characterization of deoxycytidylate methyl transferase in *Xanthomonas oryzae* infected with bacteriophyage Xp 12. Eur. J. Biochem. **87**, 29—36.

FISHER, E. F., CARUTHERS, M. H., 1979: Studies on gene control regions XII. The functional significance of *Lac* operator constitutive mutation. Nucl. Acids Res. **7**, 401—416.

FLAVELL, R. A., GROSVELD, F., BUSSLINGER, M., DE BOER, E., KIOUSSIS, D., MELLOR, A. L., GOLDEN, L., WEISS, E., HURST, J., BUD, H., BULLMAN, H., SIMPSON, E., JAMES, R., TOWNSEND, A. R. M., TAYLOR, P. M., SCHMIDT, W., FERLUNGA, J., LEBEN, L., SANTAMARIA, M., ATFIELD, G., FESTENSTEIN, H., 1982: Structure and expression of the human globin genes and murine histocompatibility antigen genes. Cold Spr. Harb. Symp. Quant. Biol. **47**, 1067—1078.

FORD, J., COCA-PRADOS, M., HSU, M.-T., 1980: Enzymatic analysis of 5-methylcytosine content in eukaryotic DNA: Study of intracellular simian virus 40 DNA. J. Biol. Chem. **255**, 7544—7547.

FRADIN, A., MANLEY, J. L., PRIVES, C. L., 1982: Methylation of simian virus 40 Hpa II site affects late, but not early, viral gene expression. Proc. Natl. Acad. Sci. U.S.A. **79**, 5142—5146.

FRIEND, C., PATULEIA, M. C., DE HARVEN, E., 1966: Erythrocytic maturation in vitro of murine (Friend) virus-induced leukemic cells. Natl. Cancer Inst. Monogr. **22**, 505.

FRITSCH, E. F., SHEN, C. K., LAWN, R. M., MANIATIS, T., 1980: The organization of repetitive sequences in mammalian globin gene clusters. Cold Spr. Harb. Symp. Quant. Biol. **45**, 761—775.

FURST, A., BROWN, E. H., BRAUNSTEIN, J. D., SCHILDKRAUT, C. L., 1981: $\beta$-Globin sequences are located in a region of early-replicating DNA in murine erythroelukemia cells. Proc. Natl. Acad. Sci. U.S.A. **78**, 1023—1027.

GEHRING, W., 1968: The stability of the determined state in cultures of imaginal disks in *Drosophila*. In: The stability of the differentiated state (URSPRUNG, H., ed.), pp. 136—154. Berlin-Heidelberg-New York: Springer.

GERBER-HUBER, S., FELBER, B. K., WEBER, R., RYFFEL, G. V., 1981: Estrogen induces tissue specific changes in the chromatin conformation of the vitellogenin genes in *Xenopus*. Nucl. Acids Res. **9**, 2475—2494.

— MAY, F. E. B., WESTLEY, B. R., FELBER, B. K., HOSBACH, H. A., ANDRES, A.-C., RYFFEL, G. V., 1983: In contrast to other *Xenopus* genes the estrogen-inducible vitellogenin genes are expressed when totally methylated. Cell **33**, 43—51.

GERMAN, J., 1962: III. DNA synthesis in human chromosomes. Trans. N.Y. Acad. Sci. **24**, 395—407.

GJERSET, R., MARTIN, D. W., 1982: Presence of a DNA demethylating activity in the nucleus of murine erythroleukemic cells. J. Biol. Chem. **257**, 8581—8583.

GLICKMAN, B., VAN DEN ELSEN, P., RADMAN, M., 1978: Induced mutagenesis in dam⁻ mutants of *Escherichia coli*: a role for 6-methyladenine residues in mutation avoidance. Mol. Gen. Genet. **163**, 307—312.

— RADMAN, M., 1980: *Escherichia coli* mutator mutants deficient in methylation instructed DNA mismatch correction. Proc. Natl. Acad. Sci. U.S.A. **77**, 1063—1067.

GOLD, M., HURWITZ, J., 1963: The enzymatic methylation of the nucleic acids. Cold Spr. Harb. Symp. Quant. Biol. **38**, 149—156.

— — ANDERS, M., 1963 a: The enzymatic methylation of RNA and DNA. Biochem. Biophys. Res. Commun. **11**, 107—114.

— — 1963 b: The enzymatic methylation of RNA and DNA: On the species specificity of the methylation enzymes. Proc. Natl. Acad. Sci. U.S.A. **50**, 164—168.

GORDON, J. W., RUDDLE, F. H., 1981: Mammalian gonadal determination and gametogenesis. Science **211**, 1265—1271.

GRIDLEY, N. D. F., 1983: Transposition of Tn3 and related transposons. Cell **32**, 3—5.

GRIPPO, P., IACCARINO, M., PARISI, E., SCARANO, E., 1968: Methylation of DNA in developing sea urchin embryos. J. Mol. Biol. **36**, 195—208.

— PARISI, E., CARESTIA, C., SCARANO, E., 1970: A novel origin of some deoxyribonucleic acid thymine and its nonrandom distribution. Biochem. **9**, 2605—2609.

GROUDINE, M., EISENMANN, R., WEINTRAUB, H., 1981: Chromatin structure of endogenous retroviral genes and activation by an inhibitor of DNA methylation. Nature **292**, 311—317.

GRUENBAUM, Y., CEDAR, H., RAZIN, A., 1982: Substrate and sequence specificity of a eukaryotic DNA methylase. Nature **295**, 620—622.

— NAVEH-MANY, T., CEDAR, H., RAZIN, A., 1981: Sequence specificity of methylation in higher plants. Nature **292**, 860—862.

— SZYF, M., CEDAR, H., RAZIN, A., 1983: Kinetics of mouse L-cell DNA methylation. (Submitted for publication.)

GRUMBACH, M. M., MORISHIMA, A., TAYLOR, J. H., 1963: Human sex chromosome abnormalities in relation to DNA replication and heterochromatinization. Proc. Natl. Acad. Sci. U.S.A. **69**, 3138—3141.

GURDON, J. B., 1962 a: The developmental capacity of nuclei taken from intestinal epithelium cells of feeding tadpoles. J. Embryol. Exp. Morphol. **10**, 622—640.

— 1962 b: Adult frogs derived from the nuclei of single somatic cells. Develop. Biol. **4**, 68—83.

— 1963: Nuclear transplantation in amphibia and the importance of stable nuclear changes in promoting cellular differentiation. Q. Rev. Biol. **38**, 54—78.

— LASKEY, R. A., 1970: The transplantation of nuclei from single culture cells into enucleate frogs' eggs. J. Embryol. Exp. Morphol. **24**, 249—255.

HADORN, E., 1963: Differenzierungsleistungen wiederholt fragmentierter Teilstücke männlicher Genitalscheiben von *Drosophila melanogaster* nach Kultur *in vivo*. Develop. Biol. **7**, 617—629.

— 1965: Problems of determination and transdetermination. Brookhaven Symp. Biol. **18**, 148—161.

HAIGH, L. S., OWENS, B. B., HELLEWELL, S., INGRAM, V. M., 1982: DNA methylation in chicken alpa-globin gene expression. Proc. Natl. Acad. Sci. U.S.A. **79**, 5332—5336.

HAMMER, S. M., RICHTER, B. S., HIRSCH, M. S., 1981: Activation and suppression of *Herpes simplex* virus in a human T lymphoid cell line. J. Immunol. **127**, 144—148.

HARBERS, K., SCHNIEKE, A., STUHLMANN, H., JÄHNER, D., JAENISCH, R., 1981: DNA methylation and gene expression: Endogenous retroviral genome becomes infectious after molecular cloning. Proc. Natl. Acad. Sci. U.S.A. **78**, 7609—7613.

HARLAND, R. M., 1982: Inheritance of DNA methylation in microinjected eggs of *Xenopus laevis*. Proc. Natl. Acad. Sci. U.S.A. **79**, 2323—2327.

HARRIS, M., 1982: Induction of thymidine kinase in enzyme-deficient Chinese hamster cells. Cell **29**, 483—492.

HATTMAN, S., 1982: DNA methyltransferase-dependent transcription of the phage Mu *mom* gene. Proc. Natl. Acad. Sci. U.S.A. **79**, 5518—5521.

HAY, R. T., DE PAMPHILIS, M. L., 1982: Initiation of SV40 DNA replication *in vivo*: Location and structure of 5' ends of DNA synthesized in the *ori* region. Cell **28**, 769—779.

HEFFRON, F., McCARTHY, B., OHTSUBO, H., OHTSUBO, E., 1979: DNA sequence analysis of the transposon Tn 3: Three genes and three sites involved in transposition of Tn 3. Cell **18**, 1153—1163.

HILLIARD, J. K., SNEIDER, T. W., 1975: Repair replication of parental DNA in synchronized cultures of Novikoff hepatoma cells. Nucl. Acids Res. **2**, 809—819.

HJELLE, B. L., PHILLIPS, J. A., III, SEEBURG, P. H., 1982: Relative levels of methylation in human growth hormone and chorionic somatomammotropin genes in expressing and non-expressing tissues. Nucl. Acids Res. **10**, 3459—3474.

HOLLIDAY, R., PUGH, J. E., 1975: DNA modification mechanisms and gene activity during development. Science **187**, 226—232.

HOSBACH, H. A., WYLER, T., WEBER, R., 1983: The *Xenopus laevis* globin gene family: chromosomal arrangement and gene structure. Cell **32**, 45—53.

HOTCHKISS, R. D., 1948: The quantitative separation of purines, pyrimidines, and nucleosides by paper chromatography. Biol. Chem. **175**, 315—332.

IIDA, S., MEYER, J., ARBER, W., 1981: Genesis and natural history of IS-mediated transposons. Cold Spr. Harb. Symp. Quant. Biol. **45**, 27—43.

IINO, T., 1961: A stabilizer of antigenic phase in *Salmonella abortusequi*. Genetics **46**, 1465—1469.

— KUTSUKAKE, K., 1980: Trans-acting genes of bacteriophage Pl and Mu mediate inversion of a specific DNA segment involved in flagellar phase variation of *Salmonella*. Cold Spr. Harb. Symp. Quant. Biol. **45**, 11—16.

JACOB, F., 1981: The possible and the actual. Seattle-London: Univ. of Washington Press.

JAHNER, D., STUHLMANN, H., STEWART, C. L., HARBERS, K., LÖHLER, J., SIMON, I., JAENISCH, R., 1982: De novo methylation and expression of retroviral genomes during mouse embryogenesis. Nature **298**, 623—628.

JELINEK, W. R., SCHMID, C. W., 1982: Repetitive sequences in eukaryotic DNA and their expression. Ann. Rev. Biochem. **51**, 813—844.

JENTSCH, S., GÜNTHERT, U., TRAUTNER, T. A., 1981: DNA methyltransferases affecting the sequence 5'CCGG. Nucl. Acids Res. **9**, 2753—2759.

JONES, P. A., TAYLOR, S. M., 1980: Cellular differentiation, cytidine analogs and DNA methylation. Cell **20**, 85—89.

— — 1981: Hemimethylated duplex DNAs prepared from 5-azacytidine-treated cells. Nucl. Acids Res. **9**, 2933—2947.

KAGI, J. H. R., NORDBERG, M., 1979: Metallothionein. Basel: Birkhäuser.

KAJIWARA, K., MUELLER, G. C., 1964: Molecular events in the reproduction of animal cells. III. Fractional synthesis of deoxyribonucleic acid with 5-bromodeoxyuridine and its effects on cloning efficiency. Biochim. Biophys. Acta **91**, 486—493.

KAPUT, J., SNEIDER, T. W., 1979: Methylation of somatic germ cell DNAs analyzed by restriction endonuclease digestion. Nucl. Acids Res. **7**, 2303—2322.

KARESS, R. E., RUBIN, G. M., 1982: A small tandem duplication is responsible for the unstable white-ivory mutation in *Drosophila*. Cell **30**, 71—79.

KARIN, M., ANDERSEN, R. D., SLATER, E., SMITH, K., HERSCHMAN, H. R., 1980: Metallothionein mRNA induction in HeLa cells in response to zinc or dexamethasone is a primary induction response. Nature **286**, 295—297.

KASTEN, M. B., GOWANS, B. J., LIEBERMAN, M. W., 1982: Methylation of deoxycytidine incorporated by excision-repair synthesis of DNA. Cell **30**, 509—516.

KELLY, T. J., jr., SMITH, H. O., 1970: A restriction enzyme from *Hemophilus influenzae*. II. Base sequence of the recognition site. J. Mol. Biol. **51**, 393—409.

KLEVECZ, R. R., KENISTON, B. A., 1975: The temporal structure of S phase. Cell **5**, 195—203.

KORBA, B. E., HAYS, J. B., 1982: Partially deficient methylation of cytosine in DNA at CCA/TGG sites stimulates genetic recombination of bacteriophage lambda. Cell **28**, 531—541.

KORNBERG, A., 1982: Supplement to DNA replication, pp. s 214—s 215. San Francisco: W. H. Freeman & Co.

KOSHLAND, M. E., 1975: Structure and function of the J-chain. Adv. Immunol. **10**, 41—69.

KRATZER, P. G., CHAPMAN, V. M., 1981: X chromosome reactivation in oocytes of *Mus caroli*. Proc. Natl. Acad. Sci. U.S.A. **78**, 3093—3097.

KUNNATH, L., LOCKER, J., 1982: Variable methylation of the ribosomal RNA genes of the rat. Nucl. Acids Res. **10**, 3877—3892.

KUO, T. T., HUANG, T. C., TENG, M. H., 1968: 5-Methylcytosine replacing cytosine in the deoxyribonucleic acid of bacterophage for *Xanthomonas oryzae*. J. Mol. Biol. **34**, 373—375.

— MANDEL, J. L., CHAMBON, P., 1979: DNA methylation: correlation with DNAse I sensitivity of chicken ovalbumin and conalbumin chromatin. Nucl. Acids Res. **7**, 2105—2113.

LEDER, P., 1982: The genetics of antibody diversity. Sci. Amer. **246**, 102—115.

LINDAHL, T., 1982: DNA repair enzymes. Ann. Rev. Biochem. **51**, 61—87.

LINN, S., ARBER, W., 1968: Host specificity of a DNA produced by *Escherichia coli*, X. *In vitro* restriction of phage FD replication form. Proc. Natl. Acad. Sci. U.S.A. **59**, 1300—1306.

LITTLE, P. F. R., 1982: Globin pseudogenes. Cell **28**, 683—684.

LOUGH, J., BISCHOFF, R., 1976: Differential sensitivity to 5-bromodeoxyuridine during the S phase of synchronized myogenic cells. Dev. Biol. **50**, 457—475.

LURIA, S., HUMAN, M. L., 1952: A non-hereditary host-induced variation of bacterial viruses. J. Bacteriol. **64**, 557—565.

LYON, M. F., 1961: Gene action in the X chromosome of the mouse (*Mus musculus* L.). Nature **190**, 372—373.

— 1971: Possible mechanisms of X chromosome inactivation. Nature New Biol. **232**, 229—232.

— 1972: X chromosome inactivation and developmental patterns in mammals. Biol. Rev. **47**, 1—35.

MACHATTIE, JACKOWSKI, J. B., 1977: Physical structure and deletion effects of the chloramphenicol resistance element Tn 9 in phage lambda. In: DNA insertion elements, plasmids, and episomes (BUKHARI, A. I., *et al.,* eds.), p. 219. Cold Spring Harbor, N.Y.: Cold Spring Harbor Laboratory.

MACLEOD, D., BIRD, A., 1982: DNase I sensitivity and methylation of active versus inactive rRNA genes in *Xenopus* species hybrids. Cell **29**, 211—218.

MAGCI, M. C., ISCOVE, N. N., ODARTCHENKO, N., 1982: Transient nature of early haematopoietic spleen colonies. Nature **295**, 527—529.

MANDEL, J. L., CHAMBON, P., 1979 a: DNA methylation: organ specific variations in the methylation pattern within and around ovalbumin and other chicken genes. Nucl. Acids Res. **7**, 2081—2103.

— — 1979 b: DNA methylation: correlation of DNase I sensitivity of chicken ovalbumin and conalbumin chromatin. Nucl. Acids Res. **7**, 2105—2113.

MANDEL, L. R., BOREK, E., 1963: The biosynthesis of methylated bases in ribonucleic acid. Biochem. **2**, 555—560.

MANN, M., SMITH, H. O., 1977: Specificity of Hpa II and Hae III DNA methylases. Nucl. Acids Res. **4**, 4211—3221.

MARCU, K. B., COOPER, M. D., 1982: New views of the immunoglobulin heavy chain switch. Nature **298**, 327—328.

MATHER, E. L., ALT, F. W., BOTHWELL, L. M., BALTIMORE, D., KOSHLAND, M. E., 1981: Expression of J chain RNA in cell lines representing different stages of B lymphocyte differentiation. Cell **23**, 369—378.

McCLELLAND, M., 1983: The effect of site specific methylation on restriction endonuclease cleavage (update). Nucl. Acids Res. **11**, 169—173.

McCLINTOCK, B., 1950: The origin and behavior of mutable loci in maize. Proc. Natl. Acad. Sci. U.S.A. **36**, 344—355.

— 1951: Chromosome organization and genic expression. Cold Spr. Harb. Symp. Quant. Biol. **16**, 13—63.

McGHEE, J. D., GINDER, G. D., 1979: Specific DNA methylation sites in the vacinity of the chicken beta-globin genes. Nature **280**, 419—420.

McKINNELL, R. C., 1978: Cloning: Nuclear transplantation in amphibia. Minneapolis: University of Minn. Press.

McKNIGHT, S. L., GAVIS, E. R., 1980: Expression of the herpes thymidine kinase gene in *Xenopus laevis* oocytes: an assay for the study of deletion mutants constructed in vitro. Nucl. Acids Res. **8**, 5931—5948.

— KINGSBURY, R., 1982: Transcriptional control signals of a eukaryotic protein coding gene. Science **217**, 316—324.

MARCU, K. B., COOPER, M. D., 1982: New views of the immunoglobulin heavy-chain switch. Nature **298**, 327—328.

MESELSON, M., YUAN, R., 1968: DNA restriction enzyme from *E. coli*. Nature **217**, 1110—1114.

MINTZ, B., 1978: Gene expression in neoplasia and differentiation (Harvey Soc. Lect. Series, 71), pp. 193—245. New York: Academic Press.

— CRONMILLER, C., 1981: METT-1: A karyotypically normal in vitro line of developmentally totipotent mouse tetratocarcinoma cells. Somatic Cell Genetics **7**, 489—505.

— — CUSTER, R. P., 1978: Somatic cell origin of teratocarcinomas. Proc. Natl. Acad. Sci. U.S.A. **75**, 2834—2828.

MOHANDAS, T., SPARKES, R. S., HELLKUHL, B., GRZESCHIK, K.-H., SHAPIRO, L. J., 1980: Expression of an X-linked gene from an inactive human X chromosome in mouse-human hybrid cells: Further evidence for the noninactivation of the sterioid sulfatase locus in man. Proc. Natl. Acad. Sci. U.S.A. **77**, 6759—6763.

— — SHAPIRO, L. J., 1981: Reactivation of an inactive human X chromosome: evidence for X inactivation by DNA methylation. Science **211**, 393—396.

MONOD, J., JACOB, F., 1961: General conclusions: Teleonomic mechanisms in cellular metabolism, growth and differentiation. Cold Spr. Harb. Symp. Quant. Biol. **24**, 389—401.

NOTHINGER, R., SCHUBIGER, G., 1966: Developmental behavior of fragments of symmetrical and asymmetrical imaginal discs of *Drosophila melanogaster* (*Diptera*). J. Embryol. Exp. Morph. **16**, 355—368.

NOWOCK, J., SIPPEL, A. E., 1982: Specific protein-DNA interaction at four sites flanking the chicken lysozyme gene. Cell **30**, 607—615.

OHNO, S., 1969: Evolution of sex chromosomes in mammals. Ann. Rev. Genet. **3**, 495—524.

— 1972: Ancient linkage groups and frozen accidents. Nature **244**, 259—262.

OHTSUBO, E., ZENILMAN, M., OHTSUBO, H., McCORMICK, M., MACHIDA, QC.,

MACHIDA, Y., 1980: Mechanism of insertion and cointegration mediated by IS 1 and Tn 3. Cold Spr. Harb. Quant. Biol. **45**, 283—296.

OTT, M.-O., SPERLING, L., CASSIO, D., LEVILLIERS, J., SALA-TREPAT, J., WEISS, M. C., 1982: Undermethylation at the 5' end of the albumin gene is necessary but not sufficient for albumin production by rat hepatoma cells in culture. Cell **30**, 825—833.

PIEKAROWICZ, A., BREZINSKI, R., 1980: Cleavage and methylation of DNA by the restriction endonuclease Hinf III isolated from *Haemophilus influenzae* Rf. J. Mol. Biol. **144**, 415—429.

POLLOCK, Y., STEIN, R., RAZIN, A., CEDAR, H., 1980: Methylation of foreign DNA sequences in eukaryotic cells. Proc. Natl. Acad. Sci. U.S.A. **77**, 6463—6467.

PROUDFOOT, N. J., MANIATIS, T., 1980: The structure of a human $\alpha$-globin pseudogene and its relationship to $\alpha$-globin gene duplication. Cell **21**, 537—544.

RADMAN, M., WAGNER, R. E., GLICKMAN, B. W., MESELSON, M., 1980: DNA methylation, mismatch correction and genetic stability. In: Progress in environmental mutagenesis (ALACENIC, M., ed.), pp. 121—130. New York: Elsevier/North-Holland Biomedical Press.

RAVETCH, J. V., HORIUCHI, K., ZINDER, N. D., 1978: Nucleotide sequence of the recognition site for the restriction-modification enzyme of *Escherichia coli* B. Proc. Natl. Acad. Sci. U.S.A. **75**, 2266—2270.

RAZIN, A., FRIEDMAN, J., 1981: DNA methylation and its possible biological roles. In: Progress in nucleic acid research and molecular biology, Vol. 23, pp. 33—52. New York: Academic Press.

— RIGGS, A. D., 1980: DNA methylation and gene function. Science **210**, 604—610.

REIF, H. J., ARBER, W., 1980: Genesis and natural history of IS-mediated transposons. Appendix II: Analysis of transposition of *IS 1-kan* and its relatives. Cold Spr. Harb. Symp. Quant. Biol. **45**, 40—44.

REIS, R. J. S., GOLDSTEIN, S., 1982: Variability of DNA methylation patterns during serial passage of human diploid fibroblasts. Proc. Natl. Acad. Sci. U.S.A. **79**, 3949—3953.

RHOADES, M. M., 1936: The effect of varying gene dosage on aleurone color in maize. J. Genet. **33**, 347—354.

— 1938: Effect of the Dt gene on the mutability of the *a,* allele in maize. Genetics **23**, 377—395.

— 1941: The genetic control of mutability in maize. Cold Spring Harb. Symp. Quant. Biol. **9**, 138—144.

— 1945: On the genetic control of mutability in maize. Proc. Natl. Acad. Sci. U.S.A. **31**, 91—95.

RIGGS, A. D., 1975: X inactivation, differentiation, and DNA methylation. Cytogenet. Cell Genet. **14**, 9—25.

— JONES, P. A., 1983: 5-methylcytosine, gene regulation and cancer. Adv. Cancer res. **39**.

ROBERTS, R. J., 1983: Restriction and modification enzymes and their recognition sequences. Nucl. Acids Res. **11**, r135—r167.

ROY, P. H., WEISBACH, A., 1975: DNA methylase from HeLa cell nuclei. Nucl. Acids Res. **2**, 1669—1684.

9

RUBIN, G. M., BROREIN, W. J., jr., DUNSMUIR, P., FLAVELL, A. J., LEWIS, R., STROBEL, E., TOOLE, J. J., YOUNG, E., 1980: Copia-like transposable elements in the *Drosophila* genome. Cold Spr. Harb. Symp. Quant. Biol. **45**, 619—628.

RUSSELL, C. J., WALKER, P. M. B., ELTON, R. A., SUBAK-SHARPE, J. H., 1976: Doublet frequency analysis of fractionated vertebrate nuclear DNA. J. Mol. Biol. **108**, 1—23.

RYFFEL, G. V., MUELLENER, D. B., WYLER, T., WAHLI, W., WEBER, R., 1981: Transcription of single-copy vitellogenin gene of *Xenopus* involves expression of middle repetitive DNA. Nature **291**, 429—431.

SAGER, R., GRABOWY, C., SANO, H., 1981: The mat-1 gene in *Chlamydomonas* regulates DNA methylation during gametogenesis. Cell **24**, 41—47.

— KITCHEN, R., 1975: Selective silencing of eukaryotic DNA. Science **189**, 426—433.

SALIM, M., MADEN, B. E. H., 1981: Nucleotide sequence of *Xenopus laevis* 18 S ribosomal RNA inferred from gene sequence. Nature **291**, 205—208.

SCARANO, E., IACCARINO, M., GRIPPO, P., PARISI, E., 1967: The heterogeneity of thymine methyl group origin in DNA pyrimidine isostichs of developing sea urchin embryos. Proc. Nat. Acad. Sci. U.S.A. **57**, 1394—1400.

— — — WINCKELMANS, D., 1965: On methylation of DNA during development of the sea urchin embryo. J. Mol. Biol. **14**, 603—607.

— TOSI, L., GRANIERI, A., 1977: Enzymatic modifications of DNA: A model for the molecular basis of cell differentiation. In: The biochemistry of adenosylmethionine, pp. 369—382. New York: Columbia Univ. Press.

SCHERER, S., DAVIS, R. W., 1980: Studies on the transposable element *Ty 1* of yeast II. Recombination and expression of *Ty 1* and adjacent sequences. Cold Spr. Harb. Symp. Quant. Biol. **45**, 584—591.

SCHMID, C. W., JELINEK, W. R., 1982: The Alu family of dispersed repetitive sequences. Science **216**, 1065—1070.

SEKIGUCHI, M., HAYAKAWA, H., MAKINO, F., TANAKA, K., OKADA, Y., 1976: A human enzyme that liberates uracil from DNA. Biochem. Biophys. Res. Commun. **73**, 293—299.

SHAPIRO, J. A., 1979; Molecular model for the transposition and replication of bacteriophage Mu and other transposable elements. Proc. Natl. Acad. Sci. **76**, 1933—1937.

SHEN, C.-K. J., MANIATIS, T., 1980: Tissue specific DNA methylation in a cluster of rabbit beta-like globin genes. Proc. Natl. Acad. Sci. U.S.A. **77**, 6634—6638.

SHOEMAKER, C., GOFF, S., GILBOA, Q., PASKIND, M., MITRA, S. W., BALTIMORE, D., 1980: Structure of cloned retroviral circular DNAs: Implications for virus integration. Cold Spr. Harb. Symp. Quant. Biol. **45**, 711—730.

SILVERMAN, M., SIEG, J., MANDEL, G., SIMON, M., 1980: Analysis of the functional components of the phase variation system. Cold Spr. Harb. Symp. Quant. Biol. **45**, 17—26.

SIMINOVITCH, L., MCCULLOCH, E. A., TILL, J. E., 1963: The distribution of colony-forming cells among spleen colonies. J. Cell. Comp. Physiol. **62**, 327—326.

SIMON, D., TISCHER, I., WAGNER, H., WERNER, E., KROGER, H., GRASSMANN, A., 1982: Effect of DNA methylation in vitro on the expression of SV 40 and VSPV.

In: Biochemistry of S-adenosylmethionine and related compounds (VSDLN, E., BORCHARDT, R. T., CREVELLING, C. R., eds.), pp. 267—273. London: Macmillan.

SMITH, H. O., WILCOX, K. W., 1970: A restriction enzyme from *Hemophilus influenzae*. I. Purification and general properties. J. Mol. Biol. **51**, 379—391.

SNEIDER, T. W., 1980: The 5'-cytosine in CCGG is methylated in two eukaryotic DNA's and Msp I is sensitive to methylation at this site. Nucl. Acids Res. **8**, 3829—3840.

— TEAGUE, W. M., ROGACHEVSKY, L. M., 1975: S-adenosylmethionine: DNA-cytosine 5-methyltransferase from a Novikoff rat hepatoma cell line. Nucl. Acids Res. **2**, 1685—1700.

STADLER, L. J., 1948: Spontaneous mutation at the R locus in Maize II. Race differences in mutation rate. Amer. Nat. **83**, 5—30.

STEIN, R., GRUENBAUM, Y., POLLACK, Y., RAZIN, A., CEDAR, H., 1982: Clonal inheritance of the pattern of DNA methylation in mouse cells. Proc. Natl. Acad. Sci. U.S.A. **79**, 61—65.

STEWART, T. A., MINTZ, B., 1981: Successive generations of mice produced from an established culture line of euploid teratocarcinoma cells. Proc. Natl. Acad. Sci. U.S.A. **78**, 6314—6318.

SUTTER, D., DOERFLER, W., 1980: Methylation of integrated adenovirus type 12 DNA sequences in transformed cells is inversely correlated with viral gene expression. Proc. Natl. Acad. Sci. U.S.A. **77**, 253—256.

— WESTFALL, M., DOERFLER, W., 1978: Patterns of integration of viral DNA sequences in the genomes of adenovirus type 12-transformed hamster cells. Cell **14**, 569—585.

TANTRAVAHI, V., GUNTAKA, R. V., ERLANGER, B. F., MILLER, O. J., 1981: Amplified ribosomal RNA genes in a rat hepatoma cell line are enriched in 5-methylcytosine. Proc. Natl. Acad. Sci. U.S.A. **78**, 489—493.

TAYLOR, J. H., 1960: Asynchronous duplication of chromosomes in cultured cells of Chinese hamster. J. Biophys. Biochem. Cytol. **7**, 455—464.

— 1977: Control of initiation of DNA replication in mammalian cells. In: DNA synthesis: present and future (MOLINEUX, I., KOHIYAMA, M., eds.). New York-London: Plenum.

— 1979: Enzymatic methylation of DNA: Patterns and possible regulatory roles. In: Molecular Genetics III, pp. 89—115. New York: Academic Press.

— 1983: DNA replication in mammalian cells. In: Molecular events in the replication of viral and cellular genomes (BECKER, Y., ed.), pp. 115—130. Boston: Martinus Nijhoff.

— WATANABE, S., 1981: Eukaryotic origins: studies of a cloned segment from *Xenopus laevis* and comparisons with human BLUR clones. In: Structure and DNA-protein interactions of replication origins (ICN-UCLA Symposium on Molecular and Cellular Biology XXI) (RAY, D. S., FOX, C. F., eds.), pp. 597—606.

TILL, J. E., 1981: Cellular diversity in the blood-forming system. Amer. Sci. **69**, 522—527.

TOSI, L., GRANIERI, A., SCARANO, E., 1972: Enzymatic DNA modifications in isolated nuclei from developing sea urchin embryos. Exp. Cell Res. **72**, 257—264.

TURNBULL, J. F., ADAMS, R. L. P., 1976: DNA methylase: purification from ascites cells and the effect of various DNA substrates on its activity. Nucl. Acid Res. **3**, 677—695.

VAN DER PLOEG, L. H. T., FLAVELL, R. A., 1980: DNA methylation in the human β-globin locus in erythroid and nonerythroid tissues. Cell **19**, 947—958.

— GRAFTEN, J., FLAVELL, R. A., 1980: A novel type of secondary modification of two CCGG residues in human gamma delta beta-globin gene locus. Nucl. Acids Res. **8**, 4563—4574.

VARDIMON, L., KRESSMANN, A., CEDAR, H., MAECHLER, M., DOERFLER, W., 1982: Expression of a cloned adenovirus gene in inhibited by in vitro methylation. Proc. Natl. Acad. Sci. U.S.A. **79**, 1073—1074.

WAALWIJK, C., FLAVELL, R. A., 1978: DNA methylation at a CCGG sequence in the large interon of the rabbit beta-globin gene: tissue-specific variations. Nucl. Acids Res. **5**, 4631—4641.

WAGNER, E. F., STEWART, T. A., MINTZ, B., 1981: The human B-globin gene and a functional viral thymidine kinase gene in developing mice. Proc. Natl. Acad. Sci. U.S.A. **78**, 5016—5020.

WAGNER, R., jr., MESELSON, M., 1976: Repair tracts in mismatched DNA heteroduplexes. Proc. Natl. Acad. Sci. U.S.A. **73**, 4135—4139.

WAHLI, W., DAVID, I. B., RYFFEL, G. V., WEBER, R., 1981: Vitellogenesis and the vitellogenin gene family. Science **212**, 298—304.

— — WYLER, T., WEBER, R., RYFFEL, G. V., 1980: Comparative analysis of the structural organization of two closely related vitellogenin genes in *X. laevis*. Cell **20**, 107—117.

WATANABE, S., TAYLOR, J. H., 1980: Cloning of an origin of DNA replication of *Xenopus laevis*. Proc. Natl. Acad. Sci. U.S.A. **77**, 5292—5296.

WEISBROD, S., 1982: Active chromatin. Nature **297**, 289—295.

WEISMANN, A., 1892: Das Keimplasma. Eine Theorie der Vererbung. Jena: G. Fischer (English translation by W. N. PARKER and H. RONNFELDT, 1915. New York: Scribner).

WIGLER, M., 1981: The inheritance of methylation patterns in vertebrates. Cell **24**, 285—286.

— LEVY, D., PERUCHO, M., 1981: The somatic replication of DNA methylation. Cell **24**, 33—40.

WILLIS, D. B., GRANOFF, A., 1980: Frog virus 3 DNA is heavily methylated at CpG sequences. Virology **107**, 250—257.

WOLF, S. F., MIGEON, B. R., 1982: Studies of X-chromosome DNA methylation in normal human-cells. Nature **295**, 667—671.

WONG, P. M. C., CLARKE, B. J., CARR, D. H., CHUI, D. H. K., 1982: Adult hemoglobins are synthesized in erythoid colonies in vitro derived from murine circulating hemopoietic progenitor cells during embryonic development. Proc. Natl. Acad. Sci. U.S.A. **79**, 2952—2956.

WOODCOCK, D. M., ADAMS, J. K., COOPER, I. A., 1982: Characteristics of enzymatic DNA methylation in cultured cells of human and hamster origin, and the effect of DNA replication inhibition. Biochim. Biophys. Acta **69**, 15—22.

WYATT, G. R., 1951: The purine and pyrimidine composition of desoxypentose nucleic acids. Biochem. J. **48**, 584—590.

YAGI, M., KOSHLAND, M. E., 1981: Expression of the J chain gene during B cell differentiation is inversely correlated with DNA methylation. Proc. Natl. Acad. Sci. U.S.A. **78**, 4807—4911.

YOUNG, M. W., SCHWARTZ, H. E., 1980: Nomadic gene families in *Drosophila*. Cold Spr. Harb. Symp. Quant. Biol. **45**, 629—640.

YOUSSOUFIAN, H., HAMMER, S. M., HIRSCH, M. S., MULDER, C., 1982: Methylation of the viral genome in an in vitro model of herpes simplex virus latency. Proc. Natl. Acad. Sci. U.S.A. **79**, 2207—2210.

YUAN, R., 1981: Structure and mechanism of multifunctional restriction endonucleases. Ann. Rev. Biochem. **50**, 285—315.

# Subject Index

Adenoviral gene, transcription inhibited by
    methylation 87
    E 2 a region, sequence 88
Adenovirus, methylation of DNA 87
Albumin gene, methylation and function
    72—73
Alu sequences 42—45
    in the human beta globin-like cluster 49
*Ascaris*, chromosome elimination 10

*Bacillus subtilis*, SP phage, restriction
    enzymes 95

*Chlamydomonas*, methylation of chloro-
    plasts 27, 29
Chromosome, fusion-breakage-bridge cycle
    18
Corn (see maize)
CpG sites, mechanism of inhibition by its
    methylation 93

Deoxycytidine and analogs 5
Determined state
    reversibility 14
    transdetermination 14
Differentiation
    insect type 12
    mechanisms of 11
    problem posed 9
    vertebrate type 14—15
*Drosophila*
    *copia*-like elements 42
    differentiation, imaginal disk 12—13
    lack of methylation of DNA 32, 117
    transposable elements 41

Enzymes
    demethylating 9, 83
    resolvase 38
    transposase 38
*Escherichia coli*
    discovery of DNA methylation 24
    effects of methylcytosine 7
    *lac* operon 6
    methylation and repair of DNA 30, 104,
    105, 107

restriction enzymes 25—26
Eukaryotic methylases, properties of 96

Genetic elements, mobile 16
    in maize 17—23
    (see also transposable elements)
Globin gene clusters 47—48
Globin genes
    effect of methylation before trans-
    formation 90
    expression in transformed cell 90
Growth hormone gene, human, methylation
    of 73—74

*Hemophilus*, restriction endonuclease 51
Hemopoietic system 46
*Herpes simplex*, thymidine kinase 88
*Herpesvirus saimiri*, methylated DNA 63

Immunoglobulins, differentiation and
    methylation 59—61
Insertion sequences 37

Lymphocytes, methylation of immuno-
    globulin genes 58—60

Maize
    linkage map, chromosome 9, 20
    variegation 21
Methylases, bacterial
    type II 94
    type I and III 95
Metallothionein gene, methylation of
    68—69
Methylation of DNA
    changes in affinity for *lac* repressor 6
    control of pattern 114—115
    de novo 8
    discovery of 6, 24
    hemoglobin genes 50—58
    inhibits transcription 87
    maintenance of 8
    position in a pyramid of controls
    117—119
    regulation of DNA repair 107—108
    role of, an overview 111—114

roles, hypotheses 27—31
roles in regulatory mechanisms 116
sensitivity to DNase I 70
stability of pattern 108—111
transcription, inhibited by methylation of 5′ flanking region 88
viral DNA 63—65
Xanthomonas phage 4
Methylation of DNA in X chromosomes, a new model 79—83
Methylation of parts of genes, primed by synthesis in vitro 92
Methylation patterns, establishment and maintenance 94
Methylcytosine
    mutational hotspot 104—107
    position in DNA sequences 7
Methylases
    inhibitors of 101
    maintenance 8
    properties of 97
Methylated bases, original discovery 4
Methylation patterns
    deletion during differentiation 100
    maintenance of 97—100
Micrococcus radiodurans, lack of methylated DNA 108
Modifications of the genome 15—16
    differentiation 12—14
Morexella, restriction endonuclease 51
Mouse, nuclear transplants 11
Mus caroli, X chromosome inactivation 77—78
Mus musculus, X chromosomes and fertility 78

Nuclear transplants 10—11

Ovalbumin gene, methylation of 70—71

Rabbit, hemoglobin genes 53
Rana pipiens, nuclear transplants 10
Restriction enzymes, inhibited by methylation 6
Restriction-modification
    in bacteria 25—26
    in eukaryotes 27
Retroviruses
    integrated 67—68
    transposable elements 42

Ribosomal genes, methylation and expression 74—75

Salmonella, regulation of flagellar protein 32—34
Somatotropin genes, methylation of 73—74
Suppression of genes, mechanisms by methylation 84—93
SV 40 virus, effect of methylation on late proteins 84—87

Thymidine kinase gene
    methylation 82
    reactivation with azacytidine 82
Transposable genetic elements 32—37
    in bacteria 33—40
    in Drosophila 41
Transposons
    in bacteria 37—40
    in eukaryotes 40—42
    in maize 22

Variegation in maize 21
Viruses, methylation of DNA 63—65
Vitellogenin genes, Xenopus 65—67

Xanthomonas, methylation of DNA in the phage XP-12 4, 29
X chromosome
    control of activation 78
    inactivation of 76—83
Xenopus
    determined state in differentiation 15
    hemoglobin genes 54—55, 58
    injection of eggs 98, 102
    injection of oocytes 84
    nuclear transplants 10
    ribosomal DNA 74
    vitellogenin genes 65—67

Yeast, transposon 40

Zea mays
    dotted (Dt) 22
    genetic map of chromosome 9, 20
    mobile genetic elements 16
    transposition of Ds 19

# Cell Biology Monographs

**Volume 1: The Lytic Compartment of Plant Cells**
By **Ph. Matile**
1975. 59 figures and 40 plates. XIII, 183 pages.
ISBN 3-211-81296-2

**Volume 2: The Golgi Apparatus**
By **W. G. Whaley**
1975. 97 figures. XII, 190 pages.
ISBN 3-211-81315-2

**Volume 3: Lysosomes: A Survey**
By **E. Holtzman**
1976. 56 figures. XI, 298 pages.
ISBN 3-211-81316-0

**Volume 4: Mitochondria: Structure, Biogenesis and Transducing Functions**
By **H. Tedeschi**
1976. 18 figures. X, 164 pages.
ISBN 3-211-81317-9

**Volume 5: Microbodies/Peroxisomen pflanzlicher Zellen**
**Morphologie, Biochemie, Funktion und Entwicklung eines Zellorganells**
Von **B. Gerhardt**
1978. 59 Abbildungen. IX, 283 Seiten.
ISBN 3-211-81436-1

**Volume 6: Patterns of Chloroplast Reproduction**
**A Developmental Approach to Protoplasmic Plant Anatomy**
By **Th. Butterfass**
1979. 28 figures. XIV, 205 pages.
ISBN 3-211-81541-4

# Cell Biology Monographs

Volume 7: **Peroxisomes and Related Particles in Animal Tissues**
By **P. Böck, R. Kramar,** and **M. Pavelka**
1980. 60 figures. XIII, 239 pages.
ISBN 3-211-81582-1

Volume 8: **Cytomorphogenesis in Plants**
Edited by **O. Kiermayer**
1981. 202 figures. X, 439 pages.
ISBN 3-211-81613-5

Volume 9: **The Protozoan Nucleus**
**Morphology and Evolution**
By **I. B. Raikov**
Translated from the 1978 Russian Edition
by N. Bobrov and M. Verkhovtseva
1982. 1 portrait and 116 figures. XV, 474 pages.
ISBN 3-211-81678-X

Volume 10: **Sialic Acids**
**Chemistry, Metabolism and Function**
Edited by **R. Schauer**
1982. 66 figures. XIX, 344 pages.
ISBN 3-211-81707-7

**Springer-Verlag Wien New York**